Integrating Monitoring and Modelling for Understanding, Predicting and Managing Sediment Dynamics

ICCE Symposium 2016

Proceedings of the International Association of Hydrological Sciences (PIAHS) Volume 375, 2017

Okehampton, United Kingdom
11 – 15 July 2017

Editors:

A. Collins
M. Stone
A. Horowitz
I. Foster

ISBN: 978-1-5108-4635-7

Printed from e-media with permission by:

Curran Associates, Inc.
57 Morehouse Lane
Red Hook, NY 12571

Some format issues inherent in the e-media version may also appear in this print version.

Printed by Curran Associates, Inc. (2017)

International Association of Hydrological Sciences
Centre for Ecology and Hydrology
Wallingford, Oxfordshire OX10 8BB
United Kingdom

Phone: 441 491 692 442
Fax: 441 491 692 448

http://iahs.info/

Additional copies of this publication are available from:

Curran Associates, Inc.
57 Morehouse Lane
Red Hook, NY 12571 USA
Phone: 845-758-0400
Fax: 845-758-2633
Email: curran@proceedings.com
Web: www.proceedings.com

Volume 375, 2017

Integrating monitoring and modelling for understanding, predicting and managing sediment dynamics

Long-term monitoring of gully erosion in Udmurt Republic, Russia

Ivan Rysin[1,2], Ivan Grigoriev[2], Mariya Zaytseva[2], Valentin Golosov[1,3], and Aidar Sharifullin[1]

[1]Department of the Landscape Ecology, Kazan Federal University, 18 Kremlyovskaya street, Kazan, 420008, Russia

[2]Department of the Environment and Ecology, Udmurt State University, 1 Universitetskaya, Izhevsk, 423064, Russia

[3]Laboratory for Soil Erosion and Fluvial Processes, Lomonosov Moscow State University, GSP-1, Leninskie Gory street, Moscow, 119991, Russia

Correspondence to: Aidar Sharifullin (luleo123@mail.ru)

Published: 3 March 2017

Abstract. This article presents results from the long term-monitoring of gully headcut retreat rates (GHRR) between 1959 and 2015 in different parts of the Udmurt Republic and is based on the use of historical aerial photographs and field observations (measuring the distance from the gully head to a fixed reference point) (Vanmaercke et al., 2016). It was determined that GHRR decreased from 2.4 to 0.3 m yr^{-1} during the 1959–1997 observation period and the 1998–2015 period, respectively. Measurements of GHRR were made once per year for most of the monitoring sites, and twice per year (after snow-melt in May, and after the rainy season, October–November) for gullies located in the eastern part of the study area that contain high proportions of arable land. 80 % of GHRR occurred during the snowmelt period (1978–1997), and decreased to 53 % since 1997. Spatial patterns of GHRR resulting from changing hydro-climatic factors for different regions of the Udmurt Republic, as a whole, were determined based on the analysis of long-term observations at 6 meteorological stations and 4 gauging stations. The main reason for decreasing GHRR appears to be due to reductions in winter frozen soil depth. The influence of stormwater runoff more clearly occurred within the east and north parts of the Vyatka-Kama interfluve, whereas higher correlations between GHRR and frozen soil depth were found for the western parts of the Republic. The most significant increases in GHRR appear to have occurred during the warm part of the year (June–July), after > 40 mm rainstorms.

1 Introduction

Gully erosion is an important soil degradation process in the southern half of the Russian Plain. The dissolution of the USSR at the end of the 20th and beginning of the 21st century led to significant changes in land use in forested areas, including the taiga and broad-leaf forests; large areas of arable land were abandoned. These changes coincided with global warming during the last decades of the 20th and early 21st centuries. Recent decades have been characterized by significant climatic changes that resulted in substantial reductions in surface water runoff in the European territory of Russia (ETR) and Western Siberia (Bazhenova et al., 1997; Dedkov, 1990; Petelko et al., 2007).

2 Materials and methods

The research area is located in the Eastern part of the Russian Plain in the southern part of the Vyatka-Kama interfluve (Fig. 1). This area of the Udmurt Republic is located south of the taiga zone. Elevations in the study area are in the range of 120–250 m a.s.l. with maximum relative elevations occurring along the river valleys. This area of the Udmurt Republic is characterized by a temperate continental climate with annual precipitation in the range of 550–600 mm; mean annual temperatures in January and July are −12.3 and +19° respectively (Perevedencev et al., 2014). The amount of arable lands in the Udmurt Republic slightly decreased from 38.5 %

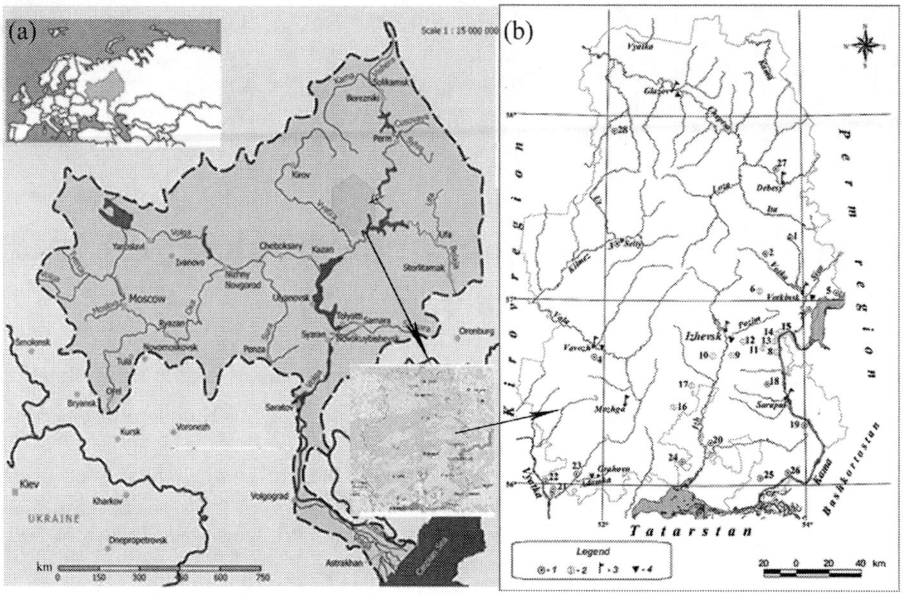

Figure 1. Map of the Volga River basin showing the location of the Vyatka-Kama interfluve **(a)** and Udmurt republic **(b)**. (Legend: 1 – study sites with annual measurement of gully head retreat, 2 – study sites with measurement of gully head retreat twice per year (after snow-melting, May; after rain-storm season, October), 3 – meteorological stations, 4 – gauging stations.

in 1955, to 35 % in 2014, with the maximum reduction occurring after the dissolution of the USSR between 1991–1996.

Gully erosion was studied by using two approaches. Firstly, high resolution aerial photographs (surveys 1959, 1970 and 1980) were used for evaluation of mean annual GHRR for the 1959–1970 and 1970–1980 periods; we used images with scales of 1 : 17 000. Gully heads were identified in all the aerial photographs and the rate of gully growth was measured (Rysin, 1998). Secondly, 168 gully heads, located at 28 sites within different parts of the study area were monitored since 1978 (Fig. 1). The actual number of monitored gully heads varied during this period because some gullies stabilized whereas new gullies became active. Different types of gullies, including slope, bank and bottom, in mostly cultivated catchments, were included in the monitoring program. GHRR was determined once per year for the majority of the sites by measuring the distance from the gully head to a fixed reference point. However, 34–40 gully heads, located in the Eastern part of the study area that contained high proportions of arable land, were measured twice per year (after snowmelt in May, and after the rainy season in October–November).

3 Results

It is important to note that the monitoring period for GHRR coincided with climate warming, which began during the second half of the 1970s (Rysin, 1998). Based on aerial photographs, the mean linear GHRR decreased from 2.4 m yr^{-1} in 1959–1970 to 1.9 m yr^{-1} in 1970–1980 for the entire study area (Fig. 2). These changes may have been due to the stabilization of the catchment area of the gully. Similar observations were noted until the mid-1990s, partly because some gullies stabilized, some arable land was abandoned during 1991–1996, and some gullies were levelled (Rysin, 1998; Rysin and Grigor'ev, 2010).

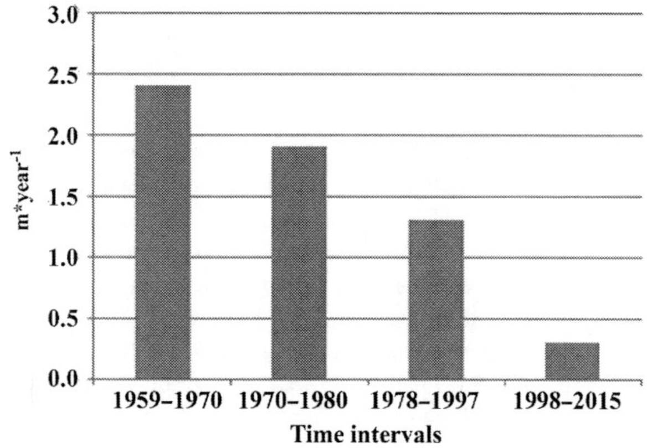

Figure 2. Mean annual gully head retreat rates for different time intervals within Udmurt Republic (time intervals 1959–1970 and 1970–1980, based on aerial photograph comparisons; time intervals 1978–1997 and 1998–2015, based on monitoring data).

There has been a considerable decrease in the mean annual rate of linear GHRR since 1997 (Fig. 3). During the 1978–2015 observation period it is possible to identify 3

Figure 3. The mean annual gully head retreat rates for the period 1978–2015, based on results of the field monitoring of gully heads at 28 sites (for site location see Fig. 1) (Legend: 1 – mean annual rates; 2 – mean rate for five-year periods).

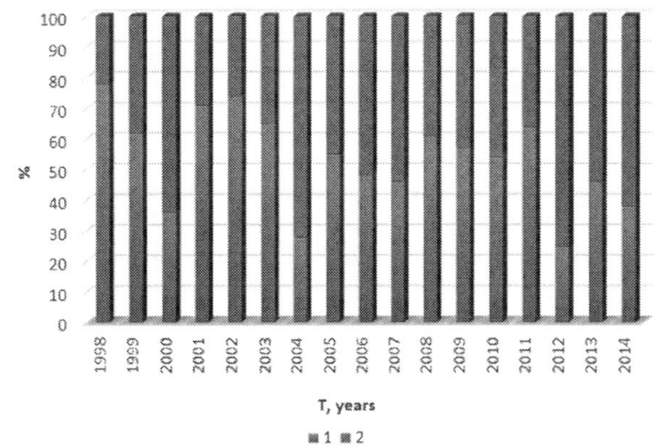

Figure 5. Annual contribution (in %) of snow-melting and rainstorms in annual gully head retreat rates for the period 1998–2014 (Legend: 1 – period of snow-melting; 2 – rain-storms period).

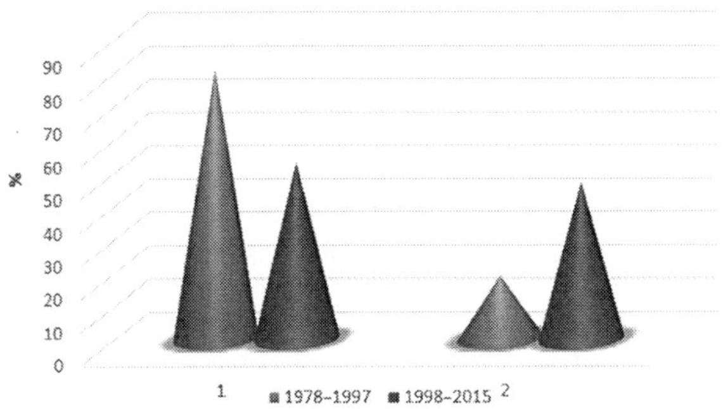

Figure 4. The mean contribution (in %) of snow-melting and rainstorms in annual gully head retreat rates for periods of monitoring 1978–1997 and 1998–2014 (Legend: 1 – period of snow-melting; 2 – rain-storms period).

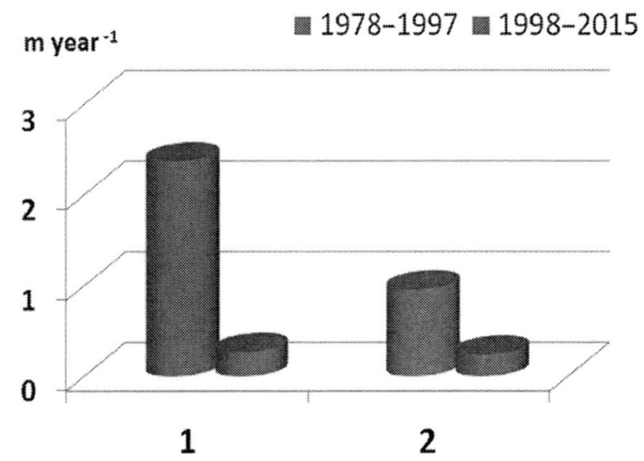

Figure 6. The mean gully head retreat for gully catchments with "warm" and "cold" aspects for the periods 1978–1997 and 1998–2015 (Legend: 1 – gully catchment with "cold" aspects (N, NE, NW and E); 2 – gully catchment with "warm" aspects (S, SW, SE, W).

peaks, with maximum values occurring in 1979 (2.8 m yr^{-1}), between 1990 and 1991 (1.9 and 2.3 m yr^{-1}), and in 1994 (1.8 m yr^{-1}). Increased rates of GHRR in these years were associated with increases in surface runoff during the spring snowmelt period (Rysin, 1998).

Based on the results of the first monitoring phase of this study (1978–1997) it is possible to conclude that 81 % of linear GHRR occurred during the snowmelt periods between March and April (Fig. 4), with the remainder occurring during the warm season. The second monitoring period (1997–2015) was characterized by a sharp decline in the mean linear GHRR (0.3 m yr^{-1}). This tendency was observed for all gully types. Further, there is a clear increasing trend in GHRR for bottom gullies since minimum values were observed in 2006.

The main reason for the substantial reductions in gullying since 1997 appears to be increasing winter air temperatures due to global warming. This led to substantial reductions in surface water runoff from the surrounding slopes during snowmelt because of an increasing frequency of years where the frozen soil depth was $<$40–50 cm. The impact of snowmelt for annual GHRR appears to have declined by some 53 %, with relatively high interannual variations (Fig. 5). It also is important to note that during the warm part of the year, serious GHRR occurred in conjunction with intense rainstorms where precipitation exceeded 40 mm. The maximum number of such rainstorms were observed within the Vyatka-Kama interfluve between 1990–1994. Since 2003, the number of extreme rainstorms ($>$40 mm) has declined in the study area (Rysin et al., 2017).

There is reason to believe that GHRR also depends on slope aspects relative to solar radiation. For the observation period, most gullies developed on the slopes that favored cold aspects (North, North–East, East, North–West; Fig. 6). In fact, the overall water resulting from snowmelt on the southern slopes is lower than on the northern slopes, where evaporation is relatively low. Also, during the second monitoring period (1998–2015), there was a noticeable reduction in GHRR for the slopes with cold aspects. The reason for this decrease probably is an increase in March temperatures during snowmelt.

Finally, it is possible to conclude that a limited number of factors influenced mean annual GHRR between 1959 and 2015, these include: land use changes, declining gully catchment areas, and climate warming. However, the latter factor appears to have led to the most notable declines in mean annual GHRR after 1996. A positive trend in mean annual GHRR has been observed for bottom gullies within the Vyatka-Kama interfluve area after 2006, and may be the initial indicator of some changes in the conditions of surface and subsurface runoff within the gullied catchments.

4 Data availability

The paper used data from a united database of gullies in the territory of the Republic of Udmurtia (Vanmaercke et al., 2016). Also we used of the database of gullies of the Udmurt Republic data, which is patented in Russia.

Competing interests. The authors declare that they have no conflict of interest.

Acknowledgements. The work was funded by Russian Scientific Fund, project no. 15-17-20006.

References

Bazhenova, O. I., Ljubcova, E. M., Ryzhov, J. V., and Makarov, S. A.: Prostranstvenno-vremennoj analiz dinamiki jerozionnyh processov na juge Vostochnoj Sibiri (Spatio-temporal analysis of the dynamics of erosion processes in the south of Eastern Siberia), Novosibirsk: Nauka (Publ.), 208 pp., 1997.

Dedkov, A. P.: Ovrazhnaja jerozija vostoka Russkoj ravniny (Gully Erosion in the East of the Russian Plain), Ed. Kazan, Kazan State University (Publ.), 142 pp., 1990.

Perevedencev, J. P., Shantalinskij, K. M., and Vazhnova, N. A.: Spatiotemporal Variations of Major Parameters of Temperature and Humidity Regime in the Volga Federal District. Meteorologija i gidrologija (Meteorology and Hydrology), 10, 19–31, 2014 (in Russian).

Petelko, A. I., Golosov, V. N., and Belyaev, V. R.: Experience of design of system of counter-erosion measures, Proceedings of the 10-th international symposium on river sedimentation, 1, 311–316, 2007.

Rysin, I. I.: Ovrazhnaja jerozija (Gully erosion), Moscow: MSU (Publ.), 168 pp., 1998.

Rysin, I. I. and Grigor'ev, I. I.: The influence of hydrometeorological agents on the ravine growth in Udmurtia. Vestnik UdGU. Serija Biologija. Nauki o Zemle (Bulletin of Udmurt University, Series Biology, Earth Sciences), 4, 137–146, 2010 (in Russian).

Rysin, I. I., Golosov, V. N., Grigoryev, I. I., and Zaitceva, M. Y.: Influence of climate change on the rates of gully growth in the Vyatka-Kama watershed // Geomorphology, 1, 90–103, 2017 (in Russian).

Vanmaercke, M., Poesen, J., Mele, B. V., Demuzere, M., Bruynseels, A., Golosov, V., Bezerra, J. F.-R, Bolysov, S., Dvinskih. A., Frankl, A., Fuseina, Yu., Guerra, A. J.-T., Haregeweyn, N., Ionita, I., Imwangana, F. M., Moeyersons, J., Moshe, I., Samani, A. N., Niacsu, L., Nyssen, J., Otsuki, Yo., Radoane, M., Rysin, I., Ryzhov, Yu. V., and Yermolaev, O.: How fast do gully headcuts retreat?, Earth-Sci. Rev., 154, 336–355, doi:10.1016/j.earscirev.2016.01.009, 2016.

Bedload transport monitoring with acoustic sensors in the Swiss Albula mountain river

Dieter Rickenmann, Gilles Antoniazza, Carlos R. Wyss, Bruno Fritschi, and Stefan Boss

Swiss Federal Research Institute WSL, Birmensdorf, 8903, Switzerland

Correspondence to: D. Rickenmann (dieter.rickenmann@wsl.ch)

Published: 3 March 2017

Abstract. Bedload transport measurements with acoustic sensors were obtained during summer 2015 in the Albula River in Switzerland. An impact plate measuring system was used with geophone and accelerometer sensors. This system provides indirect estimations of bedload transport in water courses. In April 2015, 30 impact sensors were installed in a new permanent measuring station to monitor continuously bedload transport in a mountain river with a large annual rate of sediment transport (around 90 000 $m^3\,yr^{-1}$). Records of the first year of measurement showed that (i) the signal response in terms of geophone and accelerometer impulses is comparable for both types of sensors; (ii) there is a good correlation between discharge data and impulses recorded by both types of sensors; (iii) the critical discharge at the start of bedload transport is around 5 $m^3\,s^{-1}$; (iv) a mean calibration factor for the geophone impulses can be estimated which is in a similar range as values determined from other sites with field calibration measurements.

1 Introduction

Impact measuring systems have been used successfully over the last decade to estimate bedload transport in mountain streams (Rickenmann, 2017a, b; Rickenmann et al., 2012, 2014; Reid et al., 2007; Raven et al., 2010; Beylich and Laute, 2014; Wyss et al., 2016a, b, c). As pointed out in the conclusions of the International Bedload Surrogate Monitoring Workshop in April 2007 (Gray et al., 2010) and of the International Workshop of Acoustic and Seismic Monitoring of Bedload and Mass Movements in September 2013 (Rickenmann et al., 2013), indirect impact measuring systems have the advantage of providing continuous measurements of bedload transport dynamics in both time and over a cross section whilst minimizing spatial and temporal modification of hydraulic conditions near the sensors. Nevertheless, further laboratory and field calibration experiments are needed to obtain reliable measurements of bedload transport (i.e. Rickenmann et al., 2012; Wyss et al., 2016b, c). The principle of these indirect measuring systems is to record the signals generated by the vibration induced by moving bedload particles impacting on a structure, like pipes, plates or columns (Rickenmann, 2017a, b). Impact measuring systems include the Swiss plate geophone (Rickenmann et al., 2012, 2014; Wyss et al., 2016a, b, c), the Japanese pipe hydrophone (Mizuayama et al., 2010a, b), and a wide range of other measuring devices using impact structures and different types of sensors such as geophones, accelerometers, and piezoelectric sensors (Krein et al., 2008; Reid et al., 2007; Raven et al., 2010; Tsakiris et al., 2014; Beylich and Laute, 2014).

Many previous studies using impact sensors have been conducted in small headwater catchments with high channel slopes, but relatively low annual rates of transport (i.e. Rickenmann et al., 2012, 2014; Reid et al., 2007; Raven et al., 2010). Thus, experience with impact measuring systems and bedload transport in larger alpine catchments with high annual rates of transport and bedload potential is rare, though bedload transport processes in these rivers are associated with important scientific, social and economic challenges (i.e. Badoux et al., 2014).

The Swiss plate geophone system has previously been applied successfully in many alpine streams (Rickenmann et al., 2014). Since April 2015, a new surrogate bedload transport measuring system has been operational in the Albula River in Tiefencastel, canton Grisons, Switzerland. The measurements are performed with the Swiss plate geophone system, also using accelerometer sensors in addition to the geophone sensors. The objective of this paper is to report the bedload transport monitoring during the first summer of measurements, and to compare measurements using both geophone and accelerometer sensors. The study site is of particular interest because of the relatively high annual sediment transport rate of about 90 000 $m^3\,yr^{-1}$.

2 Study site and measuring system

At the Albula River in Switzerland, the "Solis" hydropower reservoir was built in 1986. The reservoir is situated in a narrow Alpine valley, and significant sediment transport from upstream channel reaches resulted in a progressive reduction of the reservoir volume. From surveys of the sediment deposits in the hydropower lake, the annual sediment delivery has been estimated at 80 000 to 100 000 m^3. The significant annual sediment deposits in relation to the total reservoir volume resulted in a threat of the safety of the infrastructures by potentially blocking the outlets. To solve this problem, a sediment bypass tunnel was constructed in 2012, with an intake structure on the right-hand side shore of the lake and an outlet structure downstream of the concrete dam. The purpose of the bypass tunnel is to purge coarse sediments regularly to the river channel downstream of the reservoir, during periods of elevated discharges resulting in significant sediment inputs into the reservoir. The sediment bypass tunnel has a length of 970 m, a width of 4.4 m and a constant channel slope of 1.9 % (Oertli and Auel, 2015). To monitor the bedload transport during the flushing operations, a Swiss plate geophone system (Rickenmann et al., 2014) was built at the outlet of the bypass channel (Fachini et al., 2015; Hagmann et al., 2015).

The Solis reservoir is fed by the two mountain rivers; the Albula and Julia. Two artificial lakes (hydropower reservoirs) in the Julia River limit the contribution of coarse sediment into the Solis reservoir. The main input of bedload particles to the Solis reservoir therefore stems from the Albula River. Another Swiss plate geophone system to monitor bedload transport in the Albula River was built in the village of Tiefencastel in spring 2015, some 1.5 km upstream of the upper end of the Solis hydropower reservoir. The main aims of the bedload transport monitoring at this site are: (i) to develop acoustic systems further for bedload transport measurements in coarse-bedded gravel bed rivers, and (ii) to determine a more reliable sediment budget in the area of the Solis reservoir and thus to help optimize sediment management during the flushing operations.

The new bedload measuring station is located at an elevation of 855 m a.s.l., where the catchment area is around 529 km^2, of which only an insignificant portion is covered by glaciers (< 1 km^2). There are a number of water intakes related to the Solis hydropower plant, making the discharge and bedload transport partly dependent on hydroelectric operations. However, snowmelt during spring and early summer, together with rainfall events, represent the main periods of high discharges and bedload transport over the entire year.

Figure 1. Bedload measuring site in the Albula River in Tiefencastel, Switzerland. Steel plates with acoustic sensors are embedded into a sill. The width of the sill is 15 m, including a total of 30 steel plates with one sensor under each plate. Both geophones and accelerometers are installed. The view on the photos is upstream. **(a)** Low flow conditions on 16 April 2015 with a discharge of about 2 $m^3\,s^{-1}$; numbers with yellow background refer to geophone sensors, numbers with violet background to accelerometer sensors (see also Table 1); **(b)** medium flow conditions on 7 May 2015 with a discharge of about 8 $m^3\,s^{-1}$.

The channel width at the measuring cross-section in the Albula River at Tiefencastel is 15 m. A concrete sill of the same width was installed, and a steel construction was embedded into the sill. A total of 30 steel plates were mounted flush with the sill level in the steel construction (Fig. 1). The downstream side of the sill is locally slightly steeper than the bed slope upstream of the sill, which should avoid sediment deposition over the steel plates. Each steel plate is acoustically separated by elastomer elements. The steel plates are 492 mm wide, 358 mm long (in the direction of flow), and 15 mm thick.

The 20DX geophone sensors from Geospace Technologies (Houston, Texas, USA) were installed in a PC801 LPC Land-case and fixed in an aluminium box under the steel plate. Moving bedload particles impacting the steel plate cause vibrations which are transmitted by the plate to the geophone sensor. The current induced is proportional to the vibration velocity of the sensor (Rickenmann and al., 2012). The accelerometer sensor is the model KS 78.10 from Metra Mess- und Frequenztechnik (Radebeul, Germany) with a sensitivity of $10\,\mathrm{mV\,g^{-1}}$, a maximum detectable acceleration of $\pm 500\,\mathrm{g}$ and a detectable frequency range of 0.56 to 23 kHz. The measuring system includes an industrial PC sampling the signal of the acoustic sensors with a frequency of 10 kHz for both the geophone and accelerometer sensors (but including oversampling to prevent aliasing). To store the bedload transport measurements efficiently, summary values are calculated from the raw acoustic signal (Rickenmann et al., 2014) and these are stored with a resolution of one minute. Bedload transport intensity is expressed here by the number of impulses. Whenever the signal exceeds a preselected threshold value of 0.1 V (geophone sensors) or 0.4 g (accelerometer sensors), an impulse is recorded and the summed impulse counts (IMP) are stored for a given time interval of one minute.

For the first year of measurements (2015), only 23 sensors were connected to the data acquisition system (due to a limited number of available analogue-digital converter channels). During 2015, measurements were made with 15 Swiss plate geophones and 8 acceleration sensors distributed over the entire river cross-section, according to the scheme presented in Table 1.

To investigate the relationship between bedload transport and discharge at the Albula River, geophone and acceleration data were analysed together with discharge measurements. Discharge is measured at the stream gauging station of Tiefencastel, operated by the Federal Office of Environment (FOEN) and located some 500 m downstream of the acoustic bedload transport measuring station. These discharge data were corrected for bypassed water used by the hydropower company Albula-Landwasser Kraftwerke AG.

3 Preliminary results and discussion

3.1 Comparison of measurements using the two sensor types

In this section we consider data from the first "summer" season of measurements at the Albula station in Tiefencastel. For the period from 16 April until 5 July 2015, Fig. 2 presents discharge data together with the impulses summed over the 15 geophone sensors ($\mathrm{IMP_G}$). Discharge data show an increase in melt intensity from the beginning of May and discharge remains relatively high until the middle of July. Later, during the summer and autumn of 2015, discharge was mostly lower than $5\,\mathrm{m^3\,s^{-1}}$, and almost no geophone impulses were recorded. The geophone impulses correlate fairly well with discharge, and a critical discharge value corresponding to the start of bedload transport of about $5\,\mathrm{m^3\,s^{-1}}$ can be inferred from the geophone measurements.

Table 1. Numbering of the steel plates (left column) and the sensors (right column), indicting their lateral distribution; plate 1 is at river right hand side, plate 30 at river left hand side (see Fig. 1). Accelerometers in parenthesis were not connected to the data acquisition system during the measurements undertaken in 2015.

Plate	Sensor
1	Geophone 1
2	Accelerometer 1
3	Geophone 2
4	(Accelerometer 2)
5	Geophone 3
6	Accelerometer 3
7	Geophone 4
8	(Accelerometer 4)
9	Geophone 5
10	Accelerometer 5
11	Geophone 6
12	(Accelerometer 6)
13	Geophone 7
14	Accelerometer 7
15	Geophone 8
16	(Accelerometer 8)
17	Geophone 9
18	Accelerometer 9
19	Geophone 10
20	(Accelerometer 10)
21	Geophone 11
22	Accelerometer 11
23	Geophone 12
24	(Accelerometer 12)
25	Geophone 13
26	Accelerometer 13
27	Geophone 14
28	(Accelerometer 14)
29	Geophone 15
30	Accelerometer 15

Figure 3 compares discharge and number of impulses recorded by 8 acceleration sensors ($\mathrm{IMP_A}$). Generally, the trend is similar as for the geophone impulses, and again, the signal correlates fairly well with discharge. Nevertheless, some absolute and relative discrepancies can be noted. The acceleration sensors are more sensitive for recording higher frequency oscillations than the geophone sensors. In addition, the preselected threshold value for the impulse counts was found be relatively lower for the accelerometer signal (0.4 g) than for the geophone signal (0.1 V). For these reasons, the total number of impulses recorded by the accelerometers ($\mathrm{IMP_A}$) is around 31 times higher than the total number of impulses recorded by geophone ($\mathrm{IMP_G}$), if the impulse counts of the accelerometers are extrapolated to the same number of sensors as for the geophones. Furthermore, the accelerometer sensors also recorded (a relatively small

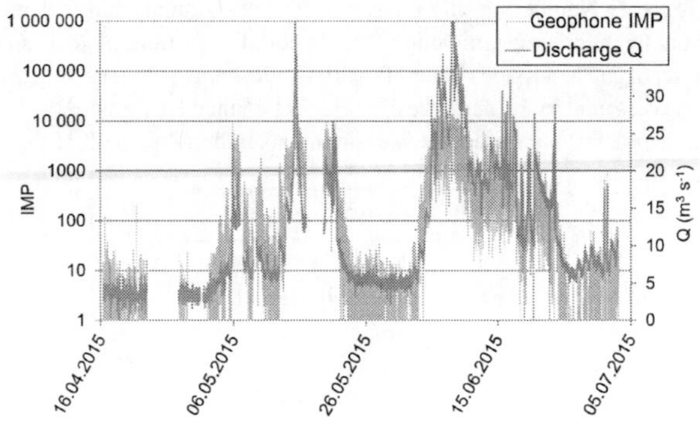

Figure 2. Discharge (Q) and number of geophone impulses (IMP_G) summed over the 15 geophone sensors, shown as a time series for the period 16 April until 5 July 2015.

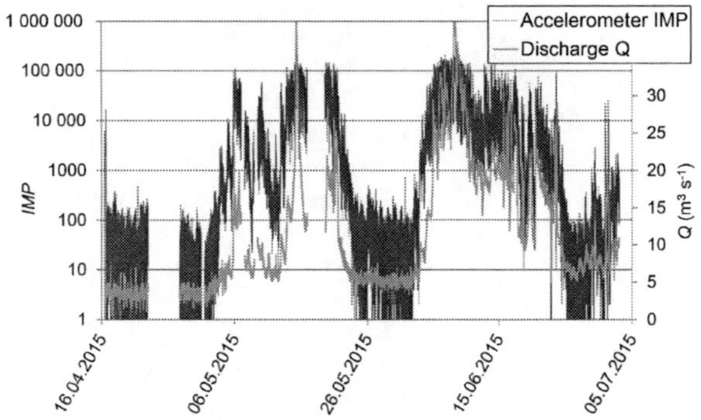

Figure 3. Discharge (Q) and number of accelerometer impulses (IMP_A) summed over 8 acceleration sensors, shown as a time series for the period 16 April until 5 July 2015.

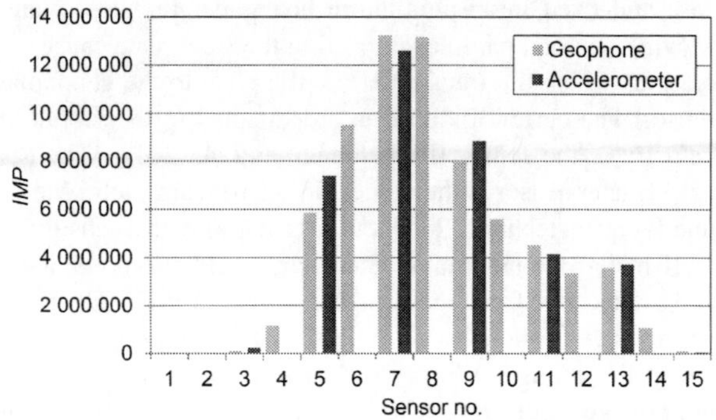

Figure 4. Total sum of impulses (IMP) recorded by each geophone sensor (yellow) and acceleration sensor (violet) during the summer of 2015. The number of impulses recorded by acceleration sensors was scaled according to Eq. (1). The view is in the upstream direction.

number) of impulses later in the season, from July until October 2015, when discharge was mostly lower than 5 $m^3\,s^{-1}$.

The sensor distribution over the cross-section allows investigation of the spatial distribution of bedload transport activity and impulse counts. Figure 4 shows the sum of impulses recorded with the two types of sensors summed over the entire season. To make the impulse counts comparable for the two sensor types, the number of impulses recorded by acceleration sensors was scaled using the following factor:

$$\alpha = (8/15) \times (\Sigma \mathrm{IMP_G} / \Sigma \mathrm{IMP_A}) \tag{1}$$

where Σ IMP_A is the total number of impulses recorded by accelerometer sensors, ΣIMP_G the total number of impulses recorded by geophone sensors over the entire season, and (8/15) the ratio between the number of recording acceleration (8) and geophone (15) sensors, respectively.

After scaling, the number of impulses recorded by the acceleration sensors appears to follow a similar pattern regarding lateral bedload transport activity as recorded by the geophone sensors. The sensors that recorded the most impulses over the season are located in the middle of the stream (sensors 7 and 8, corresponding to plate no. 13 to plate no. 15). From this location, the number of impulses recorded decreased gradually in both directions towards the river banks. However, the decline is more pronounced towards the orographic right bank (sensors 6 to 1, plates no. 11 to no. 1), than towards the orographic left bank (sensors 9 to 15, plates no. 17 to no. 30).

It appears that both the geophone and the accelerometer sensors are suitable for indirect measurements of bedload transport in gravel-bed streams. The pre-processing using simple counts of impulses (i.e. excursions of the signal above a threshold amplitude value) also works for the accelerometer sensors. The bedload transport intensity, as expressed by the impulse counts, shows a reasonable dependence on discharge.

3.2 Preliminary calibration of the Swiss plate geophone measuring system

From previous investigations with the Swiss plate geophone (SPG) system, it is known that a liner relation between the number of impulses IMP_G and mass M of bedload transported over the steel plates provides a reasonable calibration relationship (Rickenmann et al., 2014):

$$\mathrm{IMP_G} = k_b M \ [\mathrm{kg}] \tag{2}$$

where k_b [kg^{-1}] presents a linear calibration coefficient, which was found to depend on site-specific conditions.

For the Albula River measuring station, the total number of IMP_G recorded for the summer 2015 was about 7×10^7 impulses. Since only half of the bedload transport activity was monitored by the geophone sensors (as there is an accelerometer sensor under every second plate), the total number of impulses for the entire channel width is about $IMP_{G,tot} \approx 1.4 \times 10^8$. For an average year, the total volume of sediment transported into lake Solis is about 90 000 m^3, which is largely input from the Albula River (see Sect. 2). The geophone measuring site is situated about 1.2 km upstream of the upper end of the lake, and the mean channel slope in this reach is about 1.2 %. The mean channel slope over 50 m upstream of the monitoring site is about 1 %. We therefore assume that over longer time periods, the sediment transport at the measurement site is similar to the sediment input to the lake. It is known from other studies that bedload transport volumes integrated over a certain time period correlate approximately linearly with the effective runoff volume V_{re} (Rickenmann, 1997, 2001; Lenzi et al., 2004), which is the integration of the discharge above the critical discharge at initiation of transport, here taken as $25\,\mathrm{m^3\,s^{-1}}$ for the FOEN stream gauging station Tiefencastel at the Albula River, based on the acoustic measurements. The hydrologic year 2015 in Switzerland was generally quite dry. A comparison of the V_{re} values for the Albula River at Tiefencastel for the years 1987 to 2014 suggests that the V_{re} value for the year 2015 was only about 46 % of an average year. It is also known from earlier studies with the SPG system, that the impulse counts with geophone sensors only detect particles larger than about 20 mm. It is estimated that only about 20 % of the sediment deposited in (the upper part of) the lake comprise particles coarser than 20 mm, which is labelled here as BL_{20}, and it is assumed that the deposits have a porosity of about 30 % and that the grains have a density of $2650\,\mathrm{kg\,m^{-3}}$. Thus, the total mass of bedload transported past the geophone measurement site is estimated as: $\mathrm{BL}_{20} = 0.46 \times 90\,000\,\mathrm{m^3} \times 0.2 \times 0.7 \times 2650\,\mathrm{kg\,m^{-3}} = 1.5 \times 10^7\,\mathrm{kg}$. With these estimates, a mean calibration coefficient for the Albula River geophone monitoring site is estimated as $k_b \approx \mathrm{IMP}_{\mathrm{G,tot}}/\mathrm{BL}_{20} \approx 9\,[\mathrm{kg^{-1}}]$. This value is in a similar range as values determined from field calibration measurements based on direct bedload samples (Rickenmann et al., 2014).

A similar calculation can be made to arrive at a rough calibration of the accelerometer measurements. The following calibration relation is assumed to hold:

$$\mathrm{IMP}_{\mathrm{A}} = k_{\mathrm{A}} M\ [\mathrm{kg}] \tag{3}$$

where k_{A} $[\mathrm{kg^{-1}}]$ is a linear calibration coefficient. The total number of accelerometer impulses over the entire season and extrapolated to the entire channel width is about $\mathrm{IMP}_{\mathrm{A,tot}} \approx 4.3 \times 10^9$. Thus, a mean calibration coefficient for the Albula River accelerometer measurement system is estimated as $k_{\mathrm{A}} \approx \mathrm{IMP}_{\mathrm{A,tot}}/\mathrm{BL}_{20} \approx 290\,[\mathrm{kg^{-1}}]$.

4 Conclusions

An impact measuring system using geophone and accelerometer sensors was successfully installed at the Albula River (Grisons, Switzerland), an alpine mountain river with a high annual rate of sediment transport (around $90\,000\,\mathrm{m^3\,yr^{-1}}$). Preliminary results after one season of measurements showed a fairly good correlation between discharge and impulses recorded by both types of sensors, allowing the estimatation of a critical discharge corresponding to the start of bedload transport at around $5\,\mathrm{m^3\,s^{-1}}$ at the monitoring site. Most of the bedload activity occurred during the snow melt period, that is from the beginning of May until mid-July. Later, during the summer, and only for a few periods, weak bedload transport activity was recorded. Impact measuring systems also allow the investigation of the spatial distribution of bedload transport over the river cross-section. Sensors recorded a higher concentration of bedload in the middle of the stream with a gradual decrease towards the channel banks. A mean calibration coefficient for the geophone measurements was estimated based on the total number of impulses and the annual sediment delivery to the Solis lake. The calibration coefficient was found to be in a similar range to values determined previously from other sites with field calibration measurements.

5 Data availability

The data used in this study represent measurements of the first summer of recordings, and some values have to be used with caution (e.g. accelerometer impulses may include some signal noise due to the low threshold value). Those interested in using this data may contact the first author.

Competing interests. The authors declare that they have no conflict of interest.

Acknowledgements. The measuring installations and the study were supported by the Swiss Federal Office for the Environment through project "Geschiebemessungen mittels Geophonsensoren" (contract no. 06.0074.pz/m505-1377) and the Swiss Federal Research Institute WSL.

References

Badoux, A., Andres, N., and Turowski, J. M.: Damage costs due to bedload transport processes in Switzerland, Nat. Hazards Earth Syst. Sci., 14, 279–294, doi:10.5194/nhess-14-279-2014, 2014.

Beylich, A. A. and Laute, K.: Combining impact sensor field and laboratory flume measurements with other techniques for studying fluvial bedload transport in steep mountain streams, Geomorphology, 218, 72–87, 2014.

Fachini, M., Siviglia, A., and Boes, R. M.: Downstream morphological impact of a sediment bypass tunnel – preliminary results and forthcoming actions, in: Proc. First International Workshop on Sediment Bypass Tunnels, VAW-Mitteilungen 232, Laboratory of Hydraulics, Hydrology and Glaciology (VAW), ETH Zurich, Switzerland, 137–146, 2015.

Gray, J. R., Laronne, J. B., and Marr, J. D. G.: Bedload–surrogate Monitoring Technologies, U.S. Geological Survey Scientific Investigations Report 2010–5091, 37 pp., available at: http://pubs.usgs.gov/sir/2010/5091/, 2010.

Hagmann, M., Albayrak, I., and Boes, R. M.: Field research: Invert material resistance and sediment transport measurements, in: Proc. First International Workshop on Sediment Bypass Tunnels, VAW-Mitteilungen 232, Laboratory of Hydraulics, Hydrology and Glaciology (VAW), ETH Zurich, Switzerland, 123–135, 2015.

Krein, A., Klinck, H., Eiden, M., Symader, W., Bierl, R., Hoffmann, L., and Pfister, L.: Investigating the transport dynamics and the properties of bedload material with a hydro–acoustic measuring system, Earth Surf. Proc. Land., 33, 152–163, 2008.

Lenzi, M. A., Mao, L., and Comiti, F.: Magnitude-frequency analysis of bed load data in an Alpine boulder bed stream, Water Resour. Res., 40, W07201, doi:10.1029/2003WR002961, 2004.

Mizuyama, T., Oda, A., Laronne, J. B., Nonaka, M., and Matsuoka, M.: Laboratory tests of a Japanese pipe geophone for continuous acoustic monitoring of coarse bedload, in: U.S. Geological Survey Scientific Investigations Report 2010–5091, 319–335, available at: https://pubs.usgs.gov/sir/2010/5091/papers/listofpapers.html, 2010a.

Mizuyama, T., Laronne, J. B., Nonaka, M., Sawada, T., Satofuka, Y., Matsuoka, M., Yamashita, S., Sako, Y., Tamaki, S., Watari, M., Yamaguchi, S., and Tsuruta, K.: Calibration of a passive acoustic bedload monitoring system in Japanese mountain rivers, in: U.S. Geological Survey Scientific Investigations Report 2010–5091, 296–318, available at: https://pubs.usgs.gov/sir/2010/5091/papers/listofpapers.html, 2010b.

Oertli, C. and Auel, C.: Solis sediment bypass tunnel: First operation experiences, in: Proc. First International Workshop on Sediment Bypass Tunnels, VAW-Mitteilungen 232, Laboratory of Hydraulics, Hydrology and Glaciology (VAW), ETH Zurich, Switzerland, 223–233, 2015.

Raven, E. K., Lane, S. N., and Ferguson, R.: Using sediment impact sensors to improve the morphological sediment budget approach for estimating bedload transport rates, Geomorphology, 119, 125–134, 2010.

Reid, S. C., Lane, S. N., and Berney, J. M.: The timing and magnitude of coarse sediment transport events within an upland gravel-bed river, Geomorphology, 83, 152–182, 2007.

Rickenmann, D.: Sediment transport in Swiss torrents, Earth Surf. Proc. Land., 22, 937–951, 1997.

Rickenmann, D.: Comparison of bed load transport in torrents and gravel bed streams, Water Resour. Res., 37, 3295–3305, 2001.

Rickenmann, D.: Bedload transport measurements with geophones, hydrophones and underwater microphones (passive acoustic methods), in: Gravel Bed Rivers and Disasters, Wiley & Sons, Chichester, UK, 185–208, in press, 2017a.

Rickenmann, D.: Bedload transport measurements with geophones and other passive acoustic methods, J. Hydraul. Eng.-ASCE, in press, 2017b.

Rickenmann, D., Turowski, J. M., Fritschi, B., Klaiber, A., and Ludwig, A.: Bedload transport measurements at the Erlenbach stream with geophones and automated basket samplers, Earth Surf. Proc. Land., 37, 1000–1011, 2012.

Rickenmann, D., Laronne, J. B., Turowski, J. M., and Vericat, D. (Eds.): International Workshop of Acoustic and Seismic Monitoring of Bedload and Mass Movements, Abstracts, Birmensdorf, Switzerland, Swiss Federal Research Institute WSL, 109 pp., 2013.

Rickenmann, D., Turowski, J. M., Fritschi, B., Wyss, C., Laronne, J. B., Barzilai, R., Reid, I., Kreisler, A., Aigner, J., Seitz, H., and Habersack, H.: Bedload transport measurements with impact plate geophones: comparison of sensor calibration in different gravel–bed streams, Earth Surf. Proc. Land., 39, 928–942, 2014.

Tsakiris, A. G., Papanicolaou, A. N., and Lauth, T. J.: Signature of bedload particle transport mode in the acoustic signal of a geophone, J. Hydraul. Res., 52, 185–204, 2014.

Wyss, C. R., Rickenmann, D., Fritschi, B., Turowski, J. M., Weitbrecht, V., and Boes, R. M.: Measuring bedload transport rates by grain–size fraction using the Swiss plate geophone signal at the Erlenbach, J. Hydraul. Eng.-ASCE, 142, 04016003, doi:10.1061/(ASCE)HY.1943-7900.0001090, 2016a.

Wyss, C. R., Rickenmann, D., Fritschi, B., Turowski, J. M., Weitbrecht, V., and Boes, R. M.: Laboratory flume experiments with the Swiss plate geophone bedload monitoring system. Part I: Impulse counts and particle size identification, Water Resour. Res., 52, 7744–7759, doi:10.1002/2015WR018555, 2016b.

Wyss, C. R., Rickenmann, D., Fritschi, B., Turowski, J. M., Weitbrecht, V., Travaglini, E., Bardou, E., and Boes, R. M.: Laboratory flume experiments with the Swiss plate geophone bedload monitoring system. Part II: Application to field sites with direct bedload samples, Water Resour. Res., 52, 7760–7778, doi:10.1002/2016WR019283, 2016c.

The use of bed sediments in water quality studies and monitoring programs

Arthur J. Horowitz[1,2,*] **and Kent A. Elrick**[1,*]

[1]U.S. Geological Survey, Norcross, Georgia, 30093, USA
[2]Dept. of Geosciences, Georgia State University, Atlanta, Georgia, 30303, USA
[*]retired

Correspondence to: Arthur J. Horowitz (horowitz1@mindspring.com)

Published: 3 March 2017

Abstract. In most water quality monitoring programs, either filtered water (dissolved) or suspended sediment (either whole water or separated suspended sediment) are the traditional sample media of choice. This results both from regulatory requirements and a desire to maintain consistency with long-standing data collection procedures. Despite the fact that both bed sediments and/or flood plain deposits have been used to identify substantial water quality issues, they rarely are used in traditional water quality monitoring programs. The usual rationale is that bed sediment chemistry does not provide the temporal immediacy that can be obtained using more traditional sample media (e.g., suspended sediment, water). However, despite the issue of temporal immediacy, bed sediments can be used to address/identify certain types of water quality problems and could be employed more frequently for that purpose. Examples where bed sediments could be used include: (1) identifying potential long-term monitoring sites/water quality "hot spots", (2) establishing a water quality/geochemical history for a particular site/area, and (3) as a surrogate for establishing mean/median chemical values for suspended sediment.

1 Introduction

In typical environmental/water quality-monitoring programs the most common samples collected and analyzed are filtered and unfiltered (whole) water (e.g., ASTM, 2011a, b; APHA, 2012; Horowitz, 2013; USGS, various). The former are used to determine "dissolved" constituents whereas the latter are used to determine suspended sediment concentrations (SSCs)/grain-size distributions, and sediment-associated chemical constituents by subtraction. Although bed sediments have been used in large-scale (e.g., national) environmental surveys, as well as for geochemical reconnaissance (e.g., mineral exploration), they rarely are used in traditional water quality-monitoring programs (e.g., Hawkes and Webb, 1962; Webb et al., 1978; Fauth et al., 1985; Otteson et al., 2000; Gustavsson et al., 2001). However, of all the potential sample media, bed sediments probably are the easiest to collect and process, and are least likely to suffer from contamination or insufficient sample mass/analytical detection issues. Hence, their lack of use is somewhat surprising, but results from several factors. First, since the publication of the Hawkes and Webb (1962) treatise on geochemical exploration, as well as the publication of several national geochemical/environmental atlases (e.g., Webb et al., 1978; Fauth et al., 1985; Otteson et al., 2000), bed sediments have been used to detect geochemical/environmental spatial differences, but rarely are viewed as sensitive to short-term (geo)chemical variations; whereas the latter are the usual goal of most monitoring programs. Second, bed sediment surveys typically employ grain-size limited aliquots (e.g., $< 180\,\mu m$) to facilitate spatial comparisons. That traditional grain-size range has little connection with suspended sediment grain-size distributions; hence, bed sediments have never been viewed as surrogates for suspended sediment. However, recent surveys using the $< 63\,\mu m$ fraction of the upper 1 cm of bed sediments, collected after low-flow periods (e.g., Horowitz et al., 1999, 2001a, b, 2012, 2014), or surficial bed sediments collected from exposed surfaces after storms or floods (e.g., Van Metre et al., 2006; Horowitz et al., 2014), do appear to reasonably reflect suspended sed-

Table 1. Summary of Utoy Creek data from the < 63 µm fraction of bed sediment samples ($n = 47$).

Units	(mg kg^{-1})						(wt %)			
Category	Cu	Co	Ag	Cd	Pb	Zn	Al	Fe	Ti	TOC
Min	16	16	< 0.5	< 0.5	38	100	9.3	2.8	0.55	1.4
Max	120	35	0.88	0.88	200	910	12.6	6.7	0.9	5.1
Median	56.5	22	0.61	0.35	80	220	10.5	4.7	0.68	2.8
Baseline (min)*	14	8	0.1	0.2	14	71	4.9	2.2	0.25	1.3
Baseline (max)*	26	16	0.3	0.6	26	110	6.9	3.6	0.41	2.5

* From Horowitz and Stephens (2008)

iment chemical composition. Hence, grain-size limited bed sediment aliquots may represent a useful media for non-traditional monitoring programs intended to detect event-related effects, or long-term environmental/water quality-changes (e.g., decadal climate/land use-changes), provided samples are collected and analyzed every 5 to 10 years, or to determine mean/median suspended sediment associated chemical concentrations as a means of establishing first order approximations of fluvial fluxes of suspended sediment-associated chemical constituents (e.g., Horowitz et al., 2012, 2014). Additionally, the chemical analysis of cored bed sediment samples, in conjunction with absolute age dating, can be used to reconstruct the water quality/geochemical history of an impoundment (e.g., Horowitz et al., 1988, 1995; Grosbois et al., 2001; Van Metre and Horowitz, 2013; Gray et al., 2015). Three examples of water quality studies that employed bed sediments *in lieu* of more traditional sample media, as a proof of concept, are described herein.

2 Utoy Creek, Atlanta, Georgia

Utoy Creek is a small urban stream that flows through the City of Atlanta, Georgia as well as well as being a tributary of the Chattahoochee River (e.g., Horowitz and Hughes, 2006; Horowitz, 2009; West Atlanta Watershed Alliance, 2016). The creek is some 42 km long and the watershed encompasses some 89 km^2 (West Atlanta Watershed Alliance, 2016). In 2015 representatives from Fulton County contacted the authors with a request to evaluate water quality in the watershed with a view to identifying potential contaminants and to recommend long-term monitoring sites. The creek already had been listed as impaired as a result of elevated dissolved Zn levels (Andrew Mycroft, personal communication, 2015).

2.1 Methods

Bed sediment samples were manually collected during a week in late 2014 after an extended period of low-flow at roughly equidistant intervals covering the entire watershed; additionally, major tributaries to Utoy Creek also were sampled. Sampling procedures were similar to those used by Horowitz et al. (2012) and entailed collecting multiple equal volume aliquots collected from the upper 1 cm of the creek bed and/or the still wet floodplain at each site. The aliquots for each site were composited in the field, stored in clear plastic bags, and refrigerated until they could be returned to the laboratory. In total, 47 samples were collected, of which 5 were replicates to evaluate sampling precision.

All the samples were processed and analyzed following the procedures outlined in Horowitz et al. (2012). Each sample was oven-dried at 105 °C. Representative aliquots of each composite were obtained by coning and quartering the dried sediment and then wet-sieved through a 63 µm non-contaminating nylon and clear plastic sieve to obtain representative subsamples for subsequent chemical analysis. A selected group of trace elements (Ag, Al, Cd, Co, Cu, Fe, Pb, Ti, and Zn) were determined using AAS after a $HF/HClO_4$/*aqua regia* digestion in teflon beakers at 200 °C. TOC was determined by combustion using a Carlo-Erba C / N analyzer.

2.2 Results and discussion

With the exception of Cd, all the other constituents determined contained elevated individual as well as median levels that exceeded national baseline concentrations (Table 1; Horowitz and Stephens, 2008). Whilst the median concentration for Cd did not exceed national baseline levels, individual samples did.

Of particular interest to Fulton County was that whilst most of the samples contained elevated major and trace element, as well as TOC concentrations, the samples displaying the most elevated levels all clustered around one particular location in the watershed. That location is near the mouth of Utoy Creek where it discharges to the Chattahoochee River, and also is the site of a number of medium to light industries, including a galvanizing plant (Fig. 1). The survey also indicated that Utoy Creek is likely to display impaired dissolved concentrations for a number of constituents other than Zn (e.g., Cu, Co, TOC). Finally, based on this survey, potential new, long-term monitoring also sites were identified.

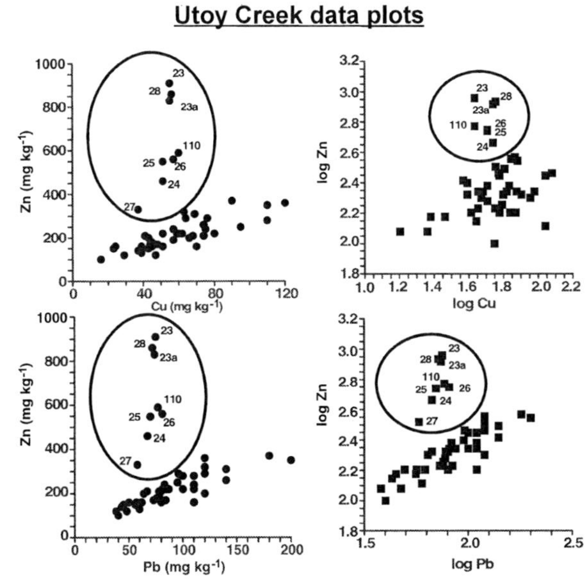

Figure 1. Plots of chemical data from the < 63 µm fraction of bed sediments from Utoy Creek, Georgia, USA. There are similar plots for Zn/Fe, Zn/TOC, Zn/Al, and Zn/Co.

3 Lake Coeur d'Alene, Idaho and the Spokane River

Lake Coeur d'Alene (CDA) is a natural (submerged river bed) lake in the northern panhandle of Idaho that formed from the outwash from Lake Missoula during the last interglacial (e.g., Hobbs et al., 1965). The lake lies between the Selkirk and the CDA Mountains and extends northward from the St. Joe River to the headwaters of the Spokane River near the city of CDA, Idaho (Meckel Engineering et al., 1983; Bender, 1991). The main body of the lake is about 3.2 km wide by 40 km long, and up to 150 m deep in the thalweg (Meckel Engineering et al., 1983). The Spokane River flows downstream from the lake, passes through a series of dams, and eventually discharges into the Columbia River, some 180 km downstream.

The South Fork of the CDA River drains a substantial part of the CDA mining district and the so-called "Silver Valley". The mining district has been in operation since the 1880's and was one of the major sources of Ag, Pb, and Zn in the US (e.g., Bender, 1991). Most of the mining and ore-processing wastes were discharged directly into the South Fork of the CDA River (e.g., Horowitz et al., 1993). As late as 1964, estimates indicated that some 2200 tonnes/day of mining and processing wastes still were entering the South Fork (Reece et al., 1978). It also has been estimated that during the course of mining, processing, and smelting operations in the area, some 115 million tonnes of mine tailings were produced and that over 60 % of this material probably entered the South Fork and the CDA River system (Javorka, 1991). These materials were highly enriched in Ag, As, Cd, Cu, Fe, Hg, Mn, Pb, Sb, and Zn (Rabe and Bauer, 1977; Bender, 1991). In 1968, tailings ponds were established to limit sediment dispersion and downstream transport (Horowitz et al., 1993). A series of studies conducted between 1989 and 1999 indicated that Lake CDA contained some 75 million tonnes of trace element rich sediment, and that elevated sediment-associated constituents could be traced all the way to the Columbia River in the state of Washington (e.g., Horowitz et al., 1993, 1995; Grosbois et al., 2001).

The local mining companies, as well as the Northwest Mining Association, contended that the majority of the im-

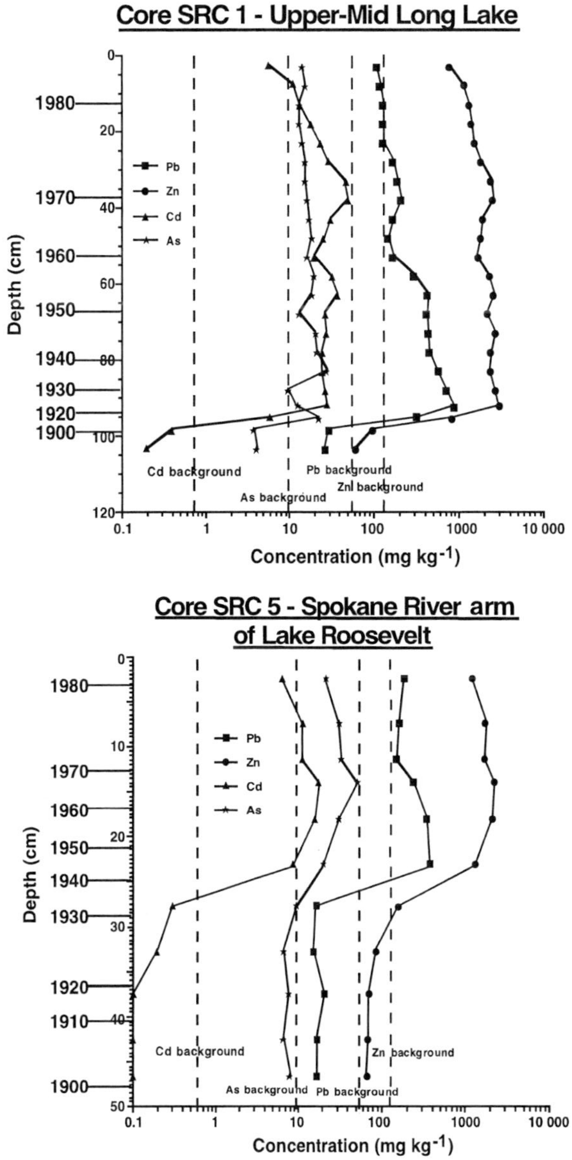

Figure 2. Plots of chemical data from two sediment cores collected in impoundments along the Spokane River, Washington, USA.

pacted sediment resulted from natural weathering processes associated with exposed segments of the ore body. To address this issue, a series of gravity cores were collected in Lake CDA and in the Spokane River system (Horowitz et al., 1995; Grosbois et al., 2001). The analytical data from the cores, in conjunction with absolute age dating based on ^{137}Cs and $^{210}Pb_{ex}$, clearly indicates that the onset of sediment-associated trace element enrichment in Lake CDA and in the Spokane River began somewhere around 1910 ± 20 years, which is roughly contemporaneous with the onset of mining and smelting operations in the region (Fig. 2). Note that the onset of sediment associated-trace element enrichment of the most downstream core that was collected in the Spokane River arm of Lake Roosevelt began around 1930 (Fig. 2; Grosbois et al., 2001). That is contemporaneous with the initial closing of the Grand Coulee Dam, downstream from the Spokane River, on the Columbia River that also caused the formation of Lake Roosevelt and the Spokane River arm (e.g., Grosbois et al., 2001). Also note that the sediment-associated trace element concentrations, at least in the Spokane River, began to decline around 1970; that is contemporaneous with the construction of the tailings ponds along the South Fork of the CDA River. This would indicate that bed sediment chemistry does appear to respond fairly quickly to land use changes and or remediation measures. The CDA ore body has been dated as Proterozoic (>500 million years BP; Hobbs et al., 1965), whereas Lake CDA has been dated as forming some 14 000 years BP (Wyman, 1993); as a result, it is highly unlikely that the vast majority of the trace element-rich sediments in Lake CDA, and in the Spokane River system, resulted from natural weathering processes.

4 The U.S. National Coastal River Survey

Between 2010 and 2011, the U.S. Geological Survey (USGS) carried out a bed sediment-associated chemical survey using material collected at or near the mouths of 132 coastal rivers in the US. The primary objective of the study was to use the chemical data generated from the samples to develop first-order approximations of the chemical fluxes emanating from the US to the coastal zone (Horowitz et al., 2012). The methods employed were essentially the same as those used in the Utoy Creek survey described earlier.

The basic premise of the study is that the chemical analyses of the <63 µm fraction of the upper 1 cm of riverbed sediments can serve as a surrogate for the average chemical composition of recent suspended sediments transiting each site. To evaluate the validity of that premise, minimum, maximum, and median suspended sediment-associated chemical concentrations determined between 1994 and 2006, as part of the revised NASQAN Program, determined for separated suspended sediment and analyzed utilizing the same procedures used in this study (Horowitz et al., 2001a, b), were compared with the chemical concentrations generated from the <63 µm fraction of the collected coastal river samples (Table 2; Horowitz et al., 2012). Based on that comparison, it appears that the premise is valid for the majority of the trace/major elements determined in the study. On the other hand, the data for carbon (TOC and TC), nitrogen (TN), and phosphorus (TP) are more ambiguous (Table 2). The bed sediment values for these constituents tend to fall either just below or just within the minimum levels for suspended sediment. This difference may be a reflection of the relatively small number of samples available for the comparison (especially for TN, and the Mississippi River Basin sites). However, the difference also may be the result of chemical/biological post-depositional remobilization that could reduce bed sediment-associated nutrient concentrations. As a result, regardless of the cause, the carbon and nutrient concentrations/annual fluxes determined from the <63 µm fraction of the bed sediments collected and analyzed during this study should be viewed as minimums.

5 Conclusions

Based on the three studies cited, it is possible to draw some conclusions about the efficacy of using bed sediment-associated major/trace element and nutrient concentrations in water quality studies and monitoring programs. It does appear as if bed sediment chemical data, particularly that generated from the <63 µm fraction of the upper 1 cm, can be used for a variety of water quality studies and monitoring programs provided that short term variations and/or temporal immediacy are not of primary concern, e.g.: for reconnaissance surveys to identify potential monitoring sites and/or "hotspots" (Utoy Creek); to establish historical reconstructions of the water quality/geochemical history of an area (Lake CDA and the Spokane River); and as surrogates for determining the mean/median concentrations of suspended sediment-associated trace/major elements and nutrients (National Coastal Survey). In the latter case, bed sediment-surrogates may provide a means of establishing long term-trends that could be associated with decadal-long processes such as climate change, without having to resort to much more resource intensive intra- and interannual sampling and analysis programs.

Data availability. Individual data are available from the author(s) on request.

Competing interests. The authors declare that they have no conflict of interest.

Table 2. Comparison between long-term suspended sediment-associated chemical levels with the < 63 µm fraction of bed sediment-associated chemical levels collected during National Coastal Survey.

Sample Name	Al	Sb	As	Ba	Be	Cd	Cr	Co	Cu	Fe	Pb	Li	Mn	Hg	Mo	Ni	TP	Se	Ag	Sr	V	Zn	Ti	TOC	TC	TN
	(%)				(mg kg^{-1})					(%)						(mg kg^{-1})								(%)		
Mississippi River @ Belle Chasse, LA*																										
n	9	9	9	9	9	9	9	9	9	9	9	9	9	9	9	9	9	9	1	9	9	9	9	6	9	9
Min.	2.4	0.4	5.6	310	0.8	0.5	40	6	21	1.3	13	24	800	0.01	2	30	890	0.5	<0.5	160	39	91	0.12	1.4	1.9	0.20
Med.	5.8	0.8	9.5	580	1.7	0.7	65	13	24	2.7	24	38	1200	0.05	2	34	1200	1.2	NA	210	83	120	0.35	2.0	2.9	0.45
Max.	6.7	0.9	11	620	2.1	2.2	73	18	27	3.4	31	46	1600	0.11	5	42	5000	2.4	0.7	360	98	150	0.44	3.2	13	1.9
Bed Sediment	6.1	0.8	11	620	1.7	0.9	60	15	23	3.0	28	37	1300	0.08	2	32	1000	0.7	<0.5	150	91	110	0.44	1.4	1.8	0.16
Wax Lake Outlet @ Calumet, LA*																										
n	9	9	9	9	9	9	9	9	9	9	9	9	9	7	9	9	9	9	3	9	9	9	9	9	9	9
Min.	2.2	0.3	4.1	360	0.6	0.3	25	4	11	1.1	7	17	370	0.04	1	13	800	0.4	<0.5	140	37	48	0.11	1.0	1.6	0.20
Med.	6.5	0.8	10	590	1.9	0.5	71	14	24	3.2	24	42	1300	0.07	2	36	1100	1.0	NA	200	95	120	0.38	2.2	2.8	0.40
Max.	7.1	0.9	12	620	2.2	0.6	79	16	37	3.5	26	54	1400	0.15	3	40	1500	1.4	0.9	410	110	150	0.41	3.6	12	0.61
Bed Sediment	6.4	0.8	10	630	1.9	0.5	62	16	23	3.3	26	41	1500	0.05	1	35	980	0.6	<0.5	140	97	120	0.44	1.3	1.7	0.14
Lower Atchafalaya River @ Morgan City, LA*																										
n	9	9	9	9	9	9	9	9	9	9	9	9	9	9	9	9	9	9	2	9	9	9	9	9	9	9
Min.	2.3	0.2	4.7	370	0.7	0.3	26	4	11	1.2	7	18	500	0.02	1	14	790	0.5	<0.5	150	37	54	0.11	0.9	1.6	0.20
Med.	6.5	0.8	11	580	1.9	0.4	69	14	23	3.4	22	43	1300	0.05	2	35	1100	1.1	NA	230	98	120	0.37	2.2	3.0	0.40
Max.	7.4	0.8	12	650	2.3	0.6	82	17	34	3.7	27	56	1600	0.11	3	42	1300	1.5	0.5	420	110	150	0.43	3.0	11	0.56
Bed Sediment	5.7	0.7	8.1	650	1.5	0.3	52	11	16	2.5	20	30	890	0.02	1	24	760	0.4	<0.5	140	77	81	0.42	0.8	1.1	0.08
Rio Grande nr Brownsville, TX																										
n	40	40	40	40	39	39	39	40	40	40	40	40	40	38	7	39	40	40	11	40	40	40	40	34	35	7
Min.	1.9	0.5	7.0	85	0.5	0.1	16	3	8	0.9	10	22	650	0.04	2	12	800	0.4	<0.5	340	27	49	0.09	1.6	3.1	0.21
Med.	5.0	1.6	10	400	1.4	0.4	37	7	17	2.3	25	44	1500	0.11	3	22	1200	0.7	NA	660	65	110	0.23	2.8	5.0	0.30
Max.	6.8	9.8	16	490	2.0	3.4	160	13	41	5.8	150	51	4100	0.42	21	88	1800	2.6	2.6	1500	90	250	0.33	4.1	9.7	0.65
Bed Sediment	5.7	0.9	7.5	480	1.7	0.3	37	8	14	2.7	19	37	490	0.05	1	18	640	0.3	<0.5	390	74	86	0.32	0.6	3.3	0.08
Colorado River @ N.I.B. above Morelos Dam nr Andrade, CA																										
n	25	25	25	25	25	21	25	25	25	25	25	25	25	16	7	25	25	25	2	25	25	25	25	14	14	2
Min.	2.0	0.3	4.4	160	0.6	0.1	36	3	8	1.1	8	21	490	0.02	6	20	400	0.3	<0.5	310	25	39	0.12	0.4	0.9	0.32
Med.	4.2	1.5	7.1	510	1.0	0.2	65	8	16	2.0	18	30	1400	0.04	8	38	700	0.9	NA	470	54	62	0.27	1.1	1.9	0.41
Max.	5.2	29	26	610	1.7	1.7	120	12	41	3.0	43	54	4500	0.26	11	76	1400	3.4	3.6	910	71	140	0.36	2.2	3.9	0.50
Bed Sediment	4.3	1.2	13	680	1.3	0.4	46	10	38	2.5	34	37	4100	0.09	2	25	980	1.2	0.5	490	65	91	0.31	1.6	3.8	0.25
Columbia River near Beaver Army Terminal, OR																										
n	101	101	101	101	101	99	95	101	101	101	101	101	101	93	31	95	101	101	8	101	101	101	101	89	97	33
Min.	5.4	0.2	2.6	380	1.0	0.1	21	9	34	2.8	7	15	590	0.02	1	15	860	0.1	<0.5	240	63	69	0.31	0.5	0.6	0.06
Med.	7.8	0.9	7.1	540	1.4	0.7	58	18	50	4.3	22	26	1100	0.07	2	36	1300	0.4	NA	320	110	160	0.50	2.2	2.3	0.32
Max.	9.5	50	19	670	2.0	1.5	200	22	150	5.6	85	100	2100	0.30	25	75	2600	1.1	1.4	450	150	340	0.66	6.1	6.2	1.1
Bed Sediment	8.0	1.0	9.6	690	1.8	1.2	61	14	47	3.4	66	26	450	0.06	<1	25	580	0.3	<0.5	320	140	220	0.65	1.5	1.6	0.14

NA – not available
* The small numbers of samples for these sites are due to a change in the most downstream sampling location for the Mississippi River from St. Francisville to Belle Chasse, and on the Atchafalaya

References

American Public Health Association (APHA): Standard Methods for the Examination of Water and Wastewater, 22nd edn., American Public Health Association: Washington, D.C. 2012, 3-6-3-7, 3-1-4-194, 3-9, 4-135-4-137, 4-107-4-108, 2012.

ASTM: Annual Book of ASTM Standards, Section 11, 11.01, Water and Environmental Technology, ASTM: West Conshohocken, Pennsylvania, 1505 pp., 2011a.

ASTM: Annual Book of ASTM Standards, Section 11, 11.02, Water and Environmental Technology, ASTM: West Conshohocken, Pennsylvania, 1507 pp., 2011b.

Bender, S.: Investigation of the chemical composition and distribution of mining wastes in Killarney Lake, Coeur d'Alene Area, northern Idaho, Unpublished MS Thesis, University of Idaho, Moscow, Idaho, 98 pp., 1991.

Fauth, H., Hindel, R., Siewers, U., and Zinner, J.: *Geochemischer Atlas Bundesrepublik Deutschland*, BGR Hannover: Schweizerbart'sche Verlagsbuchhandlung, Stuttgart, 79 pp., 1985.

Gray, J. E., Van Metre, P. C., Pribil, M. J., and Horowitz, A. J.: Tracing historical trends of Hg in the Mississippi River using Hg concentrations and Hg isotopic compositions in a lake sediment core, Lake Whittington, Mississippi, USA, Chem. Geol., 395, 80–87, 2015.

Grosbois, C., Horowitz, A. J., Smith, J. J., and Elrick, K. A.: The Effect of Mining and Related Activities on the Sediment-Trace Element Geochemistry of Lake Coeur d'Alene, Idaho, USA. Part III. Downstream Effects: The Spokane River Basin, Hydrol. Process., 15, 855–875, 2001.

Gustavsson, N., Bolviken, B., Smith, D., and Severson, R.: Geochemical landscapes of the conterminous United States – New map presentations for 22 elements, U.S. Geological Survey Professional Paper 1648, Denver, 38 pp., 2001.

Hawkes, H. and Webb, J.: *Geochemistry in Mineral Exploration*, Harper & Row, New York, 415 pp., 1962.

Hobbs, S. W., Griggs, A. B., Wallace, R. E., and Campbell, A. B.: Geology of the Coeur d'Alene District, Shoshone County, Idaho, U.S. Geological Survey Professional Paper 478, 139 pp., 1965.

Horowitz, A., Elrick, K., and Smith, J.: Annual suspended sediment and trace element fluxes in the Mississippi, Columbia, Colorado, and Rio Grande drainage basins, Hydrol. Process., 15, 1169–1207, 2001a.

Horowitz, A. and Stephens, V.: The effects of land use on fluvial sediment chemistry for the conterminous U.S. – Results from the first cycle of the NAWQA Program: Trace and major elements, phosphorus, carbon, and sulfur, Sci. Total Environ., 400, 290–314, 2008.

Horowitz, A., Elrick, K., and Smith, J.: Estimating suspended sediment and trace element fluxes in large river basins: Methodological considerations as applied to the NASQAN Program, Hydrol. Process., 15, 1107–1132, 2001b.

Horowitz, A. J.: Monitoring suspended sediments and associated chemical constituents in urban environments: Lessons from the City of Atlanta, Georgia, USA water quality monitoring program, J. Soil. Sediments, 9, 342–363, 2009.

Horowitz, A. J.: A Review of Selected Inorganic Water Quality-Monitoring Practices: Are We Really Measuring What We Think, And If So, Are We Doing It Right?, Environmental Science and Technology, 47, 2472–2486, 2013.

Horowitz, A. J. and Hughes, W. B.: The U.S. Geological Survey and City of Atlanta Water-Quality and Water-Quantity Monitoring Network, U.S. Geological Survey Fact Sheet 2005-3126, 4 pp., 2006.

Horowitz, A. J., Elrick, K. A., and Callender, E.: The effect of mining on the sediment-trace element geochemistry of cores from the Cheyenne River Arm of Lake Oahe, South Dakota, USA, Chem. Geol., 67, 17–33, 1988.

Horowitz, A. J., Elrick, K. A., and Cook, R. B.: The effect of mining and related activities on the sediment-trace element geochemistry of Lake Coeur d'Alene, Idaho, USA – Part I: Surface Sediments, Hydrol. Process., 7, 403–423, 1993.

Horowitz, A. J., Elrick, K. A., Robbins, J. A., and Cook, R. B.: The effect of mining and related activities on the sediment-trace element geochemistry of Lake Coeur d'Alene, Idaho, USA – Part II: Subsurface Sediments, Hydrol. Process., 9, 35–54, 1995.

Horowitz, A. J., Meybeck, M., Idlafkih, Z., and Biger, E.: Variations in trace element geochemistry in the Seine River Basin based on floodplain deposits and bed sediments, Hydrol. Process., 13, 1329–1340, 1999.

Horowitz, A. J., Stephens, V., Elrick, K., and Smith, J.: Concentrations and annual fluxes of sediment-associated chemical constituents from conterminous U.S. coastal rivers using bed sediment data, Hydrol. Process., 26, 1090–1114, 2012.

Horowitz, A. J., Elrick, K. A., Smith, J. J., and Stephens, V. C.: The Effects of Hurricane Irene and Tropical Storm Lee on the Bed Sediment Geochemistry of U.S. Atlantic Coastal Rivers, Hydrol. Process., 28, 1250–1259, 2014.

Javorka, E.: Lake water quality assessment Coeur d'Alene Lake, Benewah and Kootenai Counties, Idaho. Coeur d'Alene Basin Interagency Group, Coeur d'Alene Tribe of Idaho, Coeur d'Alene Subagency, Plummer, Idaho, 42 pp., 1991.

Meckel Engineering, Brown, and Caldwell Consulting Engineers: Kootenai County lakes master plan, Meckel Engineering/Brown and Caldwell, Coeur d'Alene, Idaho, 1-30-1-35, 2-24-2-32, 1983.

Otteson, R., Bogen, J., Bolviken, B., Volden, T., and Haugland, T.: Geochemical Atlas of Norway – Part 1: Chemical Composition of Overbank Sediments, Norwegian Geological Survey, Trondheim, 140 pp., 2000.

Rabe, F. and Bauer, S.: Heavy metals in lakes of the Coeur d'Alene River valley, Northwest Science, 51, 183–197, 1977.

Reece, D., Felkey, J., and Wai, C.: Heavy metal pollution in the sediments of the Coeur d'Alene River, Idaho, Environ. Geol., 2, 289–293, 1978.

U.S. Geological Survey (USGS): National field manual for the collection of water-quality data, U.S. Geological Survey Techniques of Water-Resources Investigations, Book 9, Chaps. A1–A5, available at: http://pubs.water.usgs.gov/twri9A, last access: 31 December 2016.

Van Metre, P. C. and Horowitz, A. J.: An 80-year Record of Sediment Quality in the Lower Mississippi River, Hydrol. Process., 27, 2438–2448, 2013.

Van Metre, P. C., Horowitz, A. J., Mahler, B. J., Foreman, W. T., Fuller, C. C., Burkhardt, M. R., Elrick, K. A., Furlong, E. T., Skrobialowski, S. C., Smith, J. J., Wilson, J. T., and Zaugg, S. D.: Effects of Hurricanes Katrina and Rita on the Chemistry of Bottom Sediments in Lake Ponchartrain, Louisiana, USA, Environ. Sci. Technol., 40, 6894–6902, 2006.

Webb, J., Thornton, I., Thompson, M., Howarth, R., and Lowenstein, P.: The Wolfson Geochemical Atlas of England and Wales, Oxford University Press, Oxford, 70 pp., 1978.

West Atlanta Watershed Alliance: Utoy Creek, http://wawa-online.org/our-watershed/greenspaces/utoy-creek/, last access: 25 October 2016.

Wyman, S.: The potential for heavy metal migration from sediments from Lake Coeur d'Alene in the Rathdrum Prairie Aquifer, Kootenai County, Idaho. Idaho Water Resources Research Institute, Research Technical Completion Report, University of Idaho, Moscow Idaho, 12–18, 1993.

Modelling suspended sediment distribution in the Selenga River Delta using LandSat data

Sergey R. Chalov, Varvara O. Bazilova, and Mikhail K. Tarasov

Lomonosov Moscow State University, Faculty of Geography, 119991 Leninskie gory, 1, Moscow, Russia

Correspondence to: Sergey R. Chalov (srchalov@geogr.msu.ru)

Published: 3 March 2017

Abstract. The Selenga River is the largest tributary of Baikal Lake and it's delta covers around 600 km^2. Suspended sediment concentrations (SSC) in the Selenga river delta were modelled based on LandSat images data. The seasonal variability in suspended sediment retention during the period 1989 to 2015 was calculated. The results suggest that sediment storage in the Selenga delta is observed during high discharges (> 1500 m^3 s^{-1}), whereas sediment export increases under lower flow conditions (< 1500 m^3 s^{-1}). The changes in seasonal SSC patterns are explained by wetland inundation during floods and channel erosion or Baikal wind surge during low flow periods.

1 Introduction

River mouths are important areas of sediment and chemical transport at the continental scale. They are recognized as an important lateral geochemical barrier where, over 90 % of suspended and 40 % of dissolved carbon associated with global geochemical fluxes is trapped (Lisitzyn, 1994). Sediment retention is most significant within deltas where, due to bifurcation and the existence of abundant wetlands, large volumes of sediment are stored. Deltas with complex channel networks can efficiently distribute and store sediment inputs along their channels and within floodplain environments (Syvitski et al., 2005, 2012; Rowland et al., 2009).

One of the most ancient and largest freshwater deltas in the World – the Selenga River Delta – is located at the mouth of the main tributary of the world's largest freshwater lake by volume (Lake Baikal). Due to the absence of detailed long-term and seasonal discharge and sediment monitoring, only a few studies (e.g. Potemkina, 2004) have examined sediment retention in this region. As a result, assessing the impacts of observed changes in the levels of Lake Baikal, the Selenga River water regime and sediment loads (Chalov et al., 2013, 2015) on the distribution of SSC in the Selenga River delta requires implementation of a modelling approach to provide a long-term insight into sediment and associated contaminant delivery to Lake Baikal.

2 Materials and methods

2.1 Site description

The Selenga River's 447.06 km^2 drainage basin comprises 82 % of Lake Baikal basin (Garmaev and Khristoforov, 2010). Located in the south-east end of Baikal Lake, the Selenga River delta constitutes the largest (600 km^2) wetland region in East Siberia. The Selenga River delta consists of three main distributary sectors: Lobanovskiy channel system (right, III), the Sredneustievskiy channel system (central, II), and the Selenginskiy channel system (left, I), named after the largest channels of the dispersal system (Iliycheva et al., 2015) (Fig. 1). A huge part of the delta is covered with a great number of different sized lakes. Altogether, the Selenga River Delta comprises more than 30 distributaries.

According to records from the Mostovoy gauging station (located 127 km from the delta mouth, observation period – since 1934), the average annual runoff of the Selenga is 29.3 km^3 year^{-1}. Over the period 1934–2015, the maximum observed discharge was 6420 m^3 s^{-1} (11 August 1993) and the minimum 29.9 m^3 s^{-1} (9 March 2012). The average discharges are highest during the spring-summer period with the peaks typically up to 5000–7000 m^3 s^{-1}.

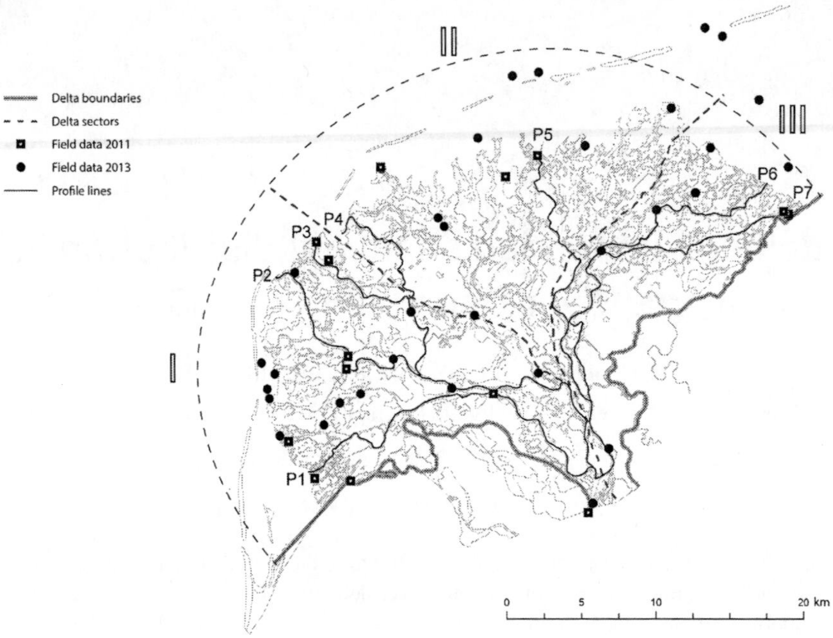

Figure 1. Selenga River delta ($P1$–$P7$ – distributaries used for ΔS calculations; I, II, III – Selenga Delta sectors; dots represent the locations of in-situ SSC measurements).

2.2 Suspended sediment concentration model based on LandSat data

This study was based on available hydrological (field campaigns with SSC measurements in 2011 and 2013) and remote sensing (82 Landsat images for the period 1989–2015) data. Field measurements of SSC for calibration of the SSC model were undertaken at more than 50 locations within the delta (Chalov et al., 2016) (Fig. 1). The available dataset was used for statistical analyses to generate least squares regression models for predicting surface SSC as a function of surface reflectance (Curran and Novo, 1988). The model relies on comparison of satellite images (raw pixel values, so called DN – digital numbers) with the SSC data obtained during field campaigns in 2011 and 2013. 2011 field measured data were compared with Landsat 5 image (17 August 2011), yielding:

$$\mathrm{SSC} = 2.418\,\mathrm{DN} - 29.91 \quad (1)$$

2013 data were compared with the Landsat 8 image (23 July 2013), yielding:

$$\mathrm{SSC} = 0.015\,\mathrm{DN} - 94.77 \quad (2)$$

The regression models (1, 2) were further applied to 82 Landsat 5 and Landsat 8 images with a spatial resolution of 30 m. The image processing approach was basically related to the calculation of SSC values ($\mathrm{mg\,L^{-1}}$) for all image pixels. An atmospheric correction method, called the "dark object method" was used for processing Landsat images. This method is based on selection of one constantly dark object on the reference image and applying the pixel value of this object to calibrate all the images in the series (Chavez, 1996). In case of the research reported here, the "dark object" was the deep part of Baikal lake. Reference images, which were used for correction comprised Landsat 5 (17 August 2011) and Landsat 8 (23 July 2013).

Since the delta consists of more than 30 branches and channels, 7 representative distributaries were selected for calculation of longitudinal change in SSC (Fig. 1). Distributaries P1, P2, P3 are located within the left channel system, P4, P5 – the central channel system, and P6, P7 – the right channel system.

The SSC balance was calculated for each distributary on the basis of the difference between SSC at the upstream (S_1) and downstream (S_2) ends:

$$\Delta S_0 = S_2 - S_1 \quad (3)$$

Values for S_1 and S_2 were estimated as an average for every 500 m of each distributary. Relative sediment retention (%) was calculated as:

$$\Delta S = \Delta S_0 / S_1 \quad (4)$$

Average ΔS values for each sector were revealed from comparison of the discharges in the upper part of the delta (Mostovoi gauging station data) (Q_o) and discharge in each channel sector (Q_i). Since there is no constantly measured discharge data in each channel system, average values for

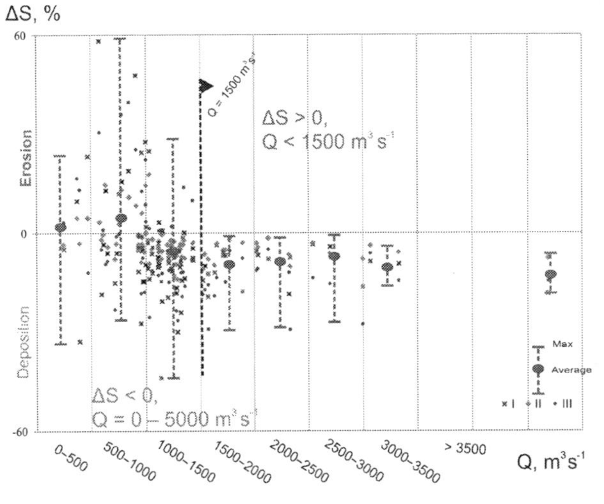

Figure 2. Suspended sediment retention (%) changes within the Selenga delta under different discharge (Q) conditions (I, II, III – delta sectors according to Fig. 1).

water partitioning between sectors (Chalov et al., 2016) were used. Three possible ratios between sediment input and output were analysed: $S_2 > S_1$, $S_2 < S_1$ and $S_2 \approx S_1$, which correspond to either erosional ($\Delta S > 0$), depositional ($\Delta S < 0$) or equilibrium ($\Delta S = 0$) states.

3 Results and discussion

Relative sediment retention (ΔS) estimates suggest that 33 % of the suspended sediment load is stored within the delta (Fig. 2). SSC ΔS is controlled by discharge. Longitudinal increases in SSC were observed only during low-water periods (discharges less than 1500 m^3 s^{-1}). In most cases, relative values of sediment retention (ΔS, %) were higher in the left or right delta sectors than in the central delta sector. Average sediment retention over 82 observational patterns was estimated at 6 % in the Left sector (I), 3 % in the Central (II) and 6 % in the Right (III) sectors. The highest deposition rate was observed in the Left delta sector ($\Delta S = 58$ %, when $Q_o = 575$ m s^{-1}). In the other delta sectors, the highest deposition rate was 17 % (II) and 39 % (III).

The observed seasonal variability of ΔS could be explained by hydrological and hydraulic factors. Longitudinal decline ($\Delta S < 0$) in water surface slopes along the delta and the corresponding order of magnitude decrease in transport capacity causes in-channel sediment storage. This is also related to the coarser bottom sediments observed in the upper part of the delta (Ilycheva et al., 2015). In addition to within channel storage, an important role in sediment retention is also played by floodplain inundation. Higher water levels increase connectivity between various delta wetlands and distributaries which causes the inundation of plants along the delta. This, in turn, causes storage of sediment on the floodplain surface and in oxbow lakes, marshes, etc. Over 50 wetland systems can be found in the lower part of the delta, during the high-discharge period, which cause significant (up to 90 %) declines in SSC along the channel systems.

The increase in sediment concentrations ($\Delta S > 0$) during the low water period is caused by the role of in-channel erosion as a component of sediment mobilization. The downstream parts of most of the delta distributaries represent actively eroded channels with high rates of bank retreat (up to 10 m) (Iliycheva et al., 2015). The probable explanation for this channel degradation which might impact on the sediment budget when the sediment delivery is relatively low (dry seasons) is that the transport capacity changes along bifurcational nodes. Under bankfull discharges, bifurcation might lead to the increase of total transport capacity of the river flow calculated as a sum of separate channels along braided (distributary) reachesd (Chalov and Alexeevsky, 2015). On the other hand, a decline in water velocity when the discharge inundates the floodplain surfaces, means that transport capacity is likely to have an opposing trend during high water periods. Another factor influencing the increase in suspended sediment transport in the lower part of the delta is the Baikal wind surge which may deliver significant amounts of sediment from Baikal into the Selenga River and then cause $\Delta S > 0$ responses.

Our preliminary results indicate hydrological drivers for the suspended sediment budget in the large distributary delta. The relationships between discharge and sediment retention rates (ΔS) could be used for predicting future delta responses. The latter is particularly important in the context of the rapid development of the Selenga River (e.g. a few dams are under construction in the Mongolian part of the catchment) which will cause changes in the hydrological regime of the delta in the near future.

4 Data availability

Underlying research data includes daily discharge from Russian HydroMet Service (accessed from AIS GMVO centre gmvo.skniivh.ru) and field data for Selenga delta (available at Supplement of Chalov et al., 2016 – see references). The 70 LandSat images were accesses from U.S. Department of the Interior U.S. Geological Survey at earthexplorer.usgs.gov.

Competing interests. The authors declare that they have no conflict of interest.

Acknowledgements. The field work was a funded by the Russian Fund for Basic Research Projects 15-05-05515 and 16-55-53116. Satellite data processing and laboratory analyses of samples were supported by the Russian Scientific Foundation project no. 14-17-00155 "River runoff parameterization for prediction of hydrological hazards and their environmental consequences".

References

Chalov, S., Jarsjö, J., Kasimov, N., Romanchenko, A., Pietron, J., Thorslund, J., and Belozerova, E.: Spatio-temporal variation of sediment transport in the Selenga River Basin, Mongolia and Russia, Environ. Earth Sci., 73, 663–680, 2015.

Chalov, S., Thorslund, J., Kasimov, N. S., Nittrouer, J., Iliyecheva, E., Pavlov, M., Pietron, J., Shinkareva, G., Lychagin, M., Aybullatov, D., Kositsky, A., Tarasov, M., Akhtman, Y., Garmaev, E., Karthe, D., and Jarsjö, J.: The Selenga River delta: a geochemical barrier protecting Lake Baikal waters, Reg. Environ. Change, doi:10.1007/s10113-016-0996-1, in press, 2016.

Chalov, S. R. and Alexeevsky, N. I.: Braided rivers: structure, types and hydrological effects, Hydrol. Res., 46, 258–275, 2015.

Chalov, S. R., Kasimov, N. S., Lychagin, M. Y., Alekseevsky, N. I., Belozerova, E. V., Theuring, P., Shinkareva, G. L., Romanchenko, A. O., and Garmaev, E. J.: Water resources assessment of the Selenga–Baikal river system, GeoÖko, 34, 77–102, 2013.

Chavez, P. S.: Image-based atmospheric corrections-revisited and improved, Photogramm. Eng. Rem. S., 62, 1025–1035, 1996.

Curran, P. J. and Novo, E. M. M.: The relationship between suspended sediment concentration and remotely sensed spectral radiance: a review, J. Coastal Res., 4, 351–368, 1988.

Garmaev, E. J. and Khristoforov, A. V.: Water resources of the rivers of the Lake Baikal Basin: basics of their use and protection, Geo, Novosibirsk, 220 pp., 2010.

Iliycheva, E. I., Gagarinova, O. V., and Pavlov, M. V.: Gidrologogeomorphologicheskiy analiz formirovaniya landshaftov v delte Selengi [Hydro-geomorphological analysis of Selenga river delta landscape formation], Geography and water resources, 3, 85–93, 2015.

Lisitzyn, A. P.: Marginal filter of oceans, Oceanology, 5, 735–747, 1994.

Potemkina, T. G.: Hydro-morphological zoning of Selenga river delta, Water Resources, 1, 15–20, 2004.

Rowland, J. C., Dietrich, W. E., Day, G., and Parker, G.: Formation and maintenance of single-thread tie channels entering floodplain lakes: Observations from three diverse river systems, J. Geophys. Res., 114, F02013, doi:10.1029/2008JF001073, 2009.

Syvitski, J. P. M., Kettner, A. J., Correggiari, A., and Nelson, B. W.: Distributary channels and their impact on sediment dispersal, Mar. Geol., 222, 75–94, doi:10.1016/j.margeo.2005.06.030, 2005.

Syvitski, J. P. M., Overeem, I., Brakenridge, G. R., and Hannon, M.: Floods, floodplains, delta plains – A satellite imaging approach, Sediment. Geol., 267–268, 1–14, doi:10.1016/j.sedgeo.2012.05.014, 2012.

Evaluation of soil erosion rates in the southern half of the Russian Plain: methodology and initial results

Valentin Golosov[1,2], Artem Gusarov[1], Leonid Litvin[2], Oleg Yermolaev[1], Nelly Chizhikova[1], Guzel Safina[1], and Zoya Kiryukhina[2]

[1]Kazan Federal University, Kremlyovskaya St. 18, Kazan, 420008, Russia
[2]Lomonosov Moscow State University, Leninskie Gory, 1, Moscow, 119991, Russia

Correspondence to: Valentin Golosov (gollossov@gmail.com) and Artem Gusarov (avgusarov@mail.ru)

Published: 3 March 2017

Abstract. The Russian Plain (RP) is divided into two principally different parts. The northern half of the RP is a predominantly forested area with a low proportion of arable fields. In contrast, the southern half of the RP has a very high proportion of arable land. During the last 30 years, this agricultural region of the RP has experienced considerable land use transformation and changes in precipitation due to climate change have altered soil erosion rates. This paper describes the use of erosion model calculations and GIS spatial analytical methods for the evaluation of trends in erosion rates in the RP. Climate change (RIHMI World Data Center, 2016), land use transformation and crop rotation modification (Rosstat, 2016; R Core Team, 2016) are the main factors governing erosion rates in the region during recent decades. It was determined that mean annual erosion rates have decreased from 7.3 to 4.1 $t\,ha^{-1}\,yr^{-1}$ in the forest zone mostly because of the serious reduction in the surface runoff coefficient for periods of snowmelt. At the same time, the erosion rates have increased from 3.9 to 4.6 $t\,ha^{-1}\,yr^{-1}$ in the steppe zone due to the increasing frequency of heavy rain-storms.

1 Introduction

The southern part of the RP has very fertile soils and is the largest agricultural region in Europe. This physiographic region consists of forest, forest-steppe and steppe landscape zones. Agricultural activity in the region has varied in duration from 4–5 centuries in the central region around Moscow in the forest zone to about 120–150 years in the south and south-east within the steppe zones (Sidorchuk and Golosov, 2003). The largest area of agricultural activity is located in the forest-steppe zone. During the 20th century, the spatial extent of agricultural activity was relatively stable within the forest and forest-steppe zone. However, it expanded in the steppe zone up to the end of the 1950s.

The last national scale quantitative assessment of erosion rates for the European part of Russia (ER) was conducted in the 1980s using a modified version of the USLE and the State Hydrological Institute model for soil erosion during periods of rain-fall and snowmelt (Larionov, 1993). Results from the model were verified using monitoring data for soil losses during snowmelt and total sedimentation in the small field ponds constructed in the early 1970's (Golosov, 2006). Estimates from the model agreed favourably with measured soil loss.

More than 35 years have passed since the last quantitative assessment of soil erosion losses from croplands on the RP. Considerable economic changes have occurred since the collapse of the USSR in 1991 and land use and crop-rotation changes in some landscape zones are evident. In addition, global warming has altered the rates and magnitude of precipitation (Groisman et al., 1999). While the effect of land use and climate change have influenced erosion rates in many parts of the World (Yang et al., 2003; Zhao et al., 2013) currently, there are no quantitative assessments of soil erosion rates in the southern portion of the predominantly agricultural RP. Because of its large spatial extent, such an assessment is only possible using a suite of remote-sensing and GIS spatial analytical methods coupled with erosion models. The methodology and initial results of this approach are presented herein to compare soil erosion rates with previous studies conducted in the 1980s.

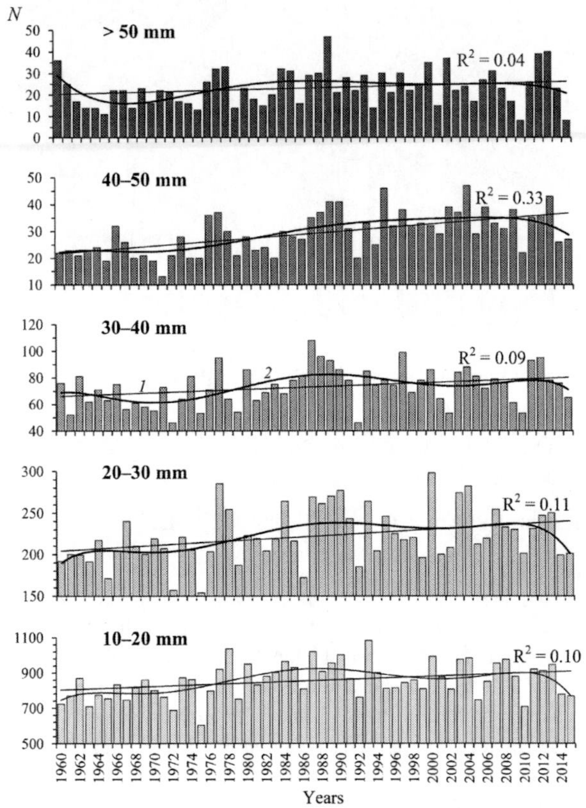

Figure 1. The mean annual number of days (N, units) with different amounts of daily precipitation (warm season, May–September) for the period 1960–2015 for all 176 meteorological stations on the RP used in this analysis. 1 – linear trend, 2 – sixth-degree polynomial trend.

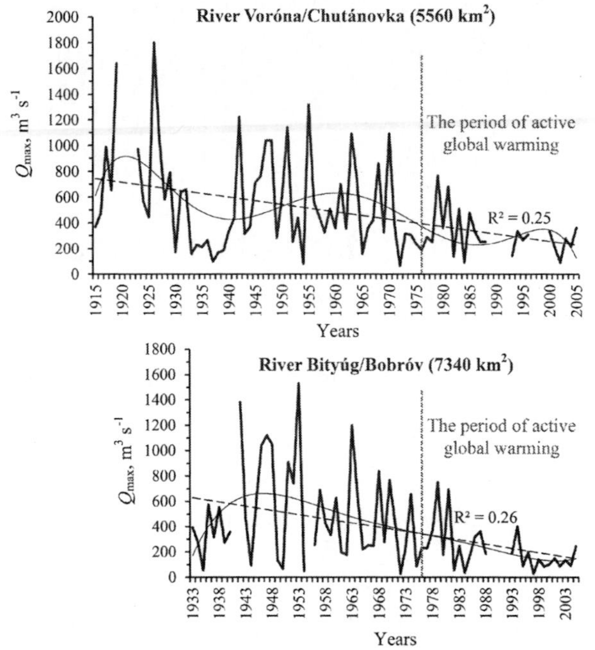

Figure 2. Trends in maximum water discharges (Q_{max}) during spring floods for some small rivers in the south of the forest-steppe zone of the RP.

2 Materials and methodology

The assessment of trends in erosion rates for the entire area of the RP is based on analysis of the dynamics of the main meteorological parameters (Dore, 2005; Zolina, 2012) and changes in both crop area and type. Estimates of present-day erosion rates for different landscapes were compared to mean annual erosion rates estimated in the 1980s.

The key meteorological parameters determining the erosion rates are soil temperature (before the period of snowmelt) and number of rain events > 10 mm. Soil temperature data for European Russia from 1960 (Park et al., 2014) was used in the models. Daily precipitation data were provided by the RIHMI World Data Center (http://aisori.meteo.ru/ClimateR; Veselov, 2002; Razuvaev et al., 1993). A total of 176 meteorological stations were selected for the analysis. The meteorological stations used in this study had precipitation data for the period 1960–2015 (several exceptions were made; a shift to 1966–2014 was accepted) and had no more than 10 % of the time series missing. According to Zolina (2012), it is appropriate to exclude stations where more than 20 % of data set is missing. Rainfall measurements were categorized as > 10, 10–20, 20–30, 30–40, 40–50 and > 50 mm by total amount and frequency. The pattern of each time series was evaluated using a generalized linear model (Madsen and Thyregod, 2012). Data processing, model building, evaluation and visualization were performed in the "R environment" (R Core Team, 2016).

Data on land-use changes and the spatial extent of land area and crop type for different regions since 1991 were provided by the Russian Federal State Statistics Service (Rosstat, 2016). Particular attention was given to evaluation of cropland dynamics in each landscape zone. Additional sources of information for correction of possible errors in the official statistics were found elsewhere (Lyuri et al., 2010). Crop rotation coefficients for modelling were calculated separately for both rainfall season and snow melt for the period 1991–2014.

3 Results

There is widespread agreement in the literature that climate change has had a significant effect on weather patterns (Dore, 2005) and erosion rates globally (Nearing et al., 2004). The total number of rain events of > 10 mm have not increased significantly for all groups since the 1960s for most parts of the RP (Fig. 1). There was, however, a significant increase

Table 1. Comparison of mean annual soil erosion rates and total soil losses between1980 and 2012.

Landscape zones (subzones) of the European part of Russia (without mountain areas)	Mean annual soil erosion rates in 1980, $t\,ha^{-1}\,yr^{-1}$	Mean annual soil erosion rates in 2012, $t\,ha^{-1}\,yr^{-1}$	Change in mean annual soil erosion rates*, (±) %	Total soil losses in 1980, 10^3 t	Total soil losses in 2012, 10^3 t	Change in total soil losses* (±) %
Northern and middle taiga subzones	6.5	4.0	−38.4	6131.5	1809	−70.5
Southern taiga subzone	7.3	4.1	−44.0	145 032	35 966	−75.2
Forest zones, total	7.3	4.1	−43.8	151 163	37 791	−75.0
Forest-steppe zone	4.1	3.3	−19.4	136 450	79 277	−41.9
Steppe zone	3.9	4.63	18.7	148 618	127 663	−14.1
Total European Russia	4.7	4.0	−15.0	436 231	244 831	−43.9

* relationship between 1980 and 2012

Table 2. Changes in the areas of arable and cultivated lands in the landscape zones of the RP between 1980 and 2012.

				% Change in area	
Landscape zones of the European part of Russia	Area of cultivated lands in 1980*, 10^3 ha	Arable lands in 2012, 10^3 ha	Area of cultivated lands in 2012, 10^3 ha	Arable lands in 2012, (±) %	Area of cultivated lands in 2012, (±) %
Forest zones (total)	20 789.9	16 738	9222.1	−19.5	−55.6
Forest-steppe zone	33 286.4	29 360	23 978.1	−11.8	−28
Steppe zone	38 544.8	36 104	27 894.3	−6.3	−27.6
Total European Russia	92 621.2	82 202	61 094.5	−11.2	−34

* Area of cultivated lands in 1980 was equal to area of arable lands

($p>0.10$) in the number of rain events >40–50 mm mostly along the line linking Bryansk–Izhevsk and near the Caucasus foothills in the south-west parts of the steppe zone. It has previously been reported that about 80 % of total soil losses occur during the warm season (Edwards and Owens, 1991) and the observed increase in the total number of heavy rains will have had an effect on the erosion in the steppe zone of the RP (Table 1).

The average soil temperature increased over the period 1960 to 2006 by 8 °C in the south-western part of the RP in the steppe zone and 4 °C in the southern part of the forest zone near the Ural mountain system (Park et al., 2014). Increasing soil temperature reduced the depth of frozen soil before snowmelt and surface runoff coefficients decreased as a consequence. This is also confirmed by a reduction in spring flood levels for most small rivers in the southern half of the RP since the beginning of the 1980s (Fig. 2). Results of direct monitoring of surface runoff from cultivated land at the Novosil' monitoring station, located in the middle part of the Zhusha River basin (Central Russia), confirmed that the coefficient of surface runoff had decreased from 0.5 for the period 1955–1980 to <0.1 for the most recent decades (Petelko et al., 2007). The observed decrease in gully head retreat rates during the last decade in different parts of the Vyatsko-Kamskoe interfluve area may also account for the reduction of surface runoff from the cultivated slopes during periods of snowmelt (Rysin et al., this volume). Accordingly, soil losses from croplands during snowmelt periods have decreased significantly over the last two decades in particular in the forest zone.

During the period 1980–2012, there was a significant reduction in cropland area due to an increase in abandoned lands that have not been cultivated for the past decade for the all landscape zones of the RP (Table 2). However, the extent of cultivated lands has increased in the forest-steppe zone since 2010.

Although some changes in crop rotation were observed, the crop and cover management coefficients have not changed for the forest-steppe and steppe zones. However, crop and cover management coefficients have decreased in the forest zone. The primary reason for this observation was an increasing proportion of cultivated land under annual and/or perennial grasses in the forest landscape zone.

Meteorological and land use change data were used to recalculate the mean annual erosion rates and the annual soil losses for the different landscape zones of the RP (Table 1). It was found that the annual soil losses had decreased in all the landscape zones because of a reduction in cropland area.

Further analyses of the factors affecting soil erosion rates at a range of spatial scales and for different landscape types is required to provide more detail regarding the influence of land use and climate change on the erosion rates in the agriculturally intensive southern portion of the RP.

4 Conclusions

Quantitative assessment of contemporary erosion rates on the arable lands of the RP was undertaken using remote sensing and GIS techniques coupled with erosion models. Temporal changes of some key parameters in the erosion model (erosion index of precipitation, crop and cover management coefficients and surface runoff coefficient for periods of snowmelt) were investigated. During the period from 1980 to 2012, mean annual erosion rates decreased from 7.3 to 4.1 $\mathrm{t\,ha^{-1}\,yr^{-1}}$ in the forest zone, because of a reduction in surface runoff coefficients during snowmelt and the increasing area under annual and perennial grasses. There was a slight decrease in erosion rates in the forest-steppe zone and a more significant increase in erosion rates in the steppe zone (Table 1). The latter change occurred because of the increasing frequency of heavy rainstorms. The reduction in soil loss since the 1980s in all landscape zones was attributed to the decrease in cropped area (Table 2).

5 Data availability

Information about arable and cultivated lands was collected and areas under different crops (information was used for calculation of crop and cover management coefficients), were taken from the official site of Rosstat – Russian Federal Service of State Statistics, Moscow, Russia, available at: http://www.gks.ru (Rosstat, 2016). Additional sources of information for correction of possible errors in the official statistics for arable and cultivated lands were found in Lyuri et al. (2010). Meteorological information about daily precipitation data were provided by the RIHMI World Data Center (2016, http://aisori.meteo.ru/ClimateR). Information about soil characteristics and LS factor (relief factor), which are also requested for soil erosion rate calculation, were collected for the initial calculation of soil losses on period 1980 and it is only available in the form of paper worksheets. In the calculations of soil erosion rates, it was assumed that these parameters remain unchanged since 1980 until 2012. It is planned to consider possible changes to the relief factor due to changes in arable areas. Information about maximum water discharge were collected from Hydrological Yearbooks, which are available only in printed version and they are not publicly accessible since 1991, because of they collected by Hydrometeorological Service of Russian Federation and concentrated in State Hydrological Institute. It is possible to receive access to them only by official request.

Competing interests. The authors declare that they have no conflict of interest.

Acknowledgements. This work was funded by the Russian Science Foundation, project no. 15–17–20006.

References

Dore, M. H. I.: Climate change and changes in global precipitation patterns: what do we know?, Environ. Int., 31, 1167–1181, doi:10.1016/j.envint.2005.03.004, 2005.

Edwards, W. M. and Owens, I. B.: Large storm effects on the total soil erosion, J. Soil Water Conserv., 46, 75–78, 1991.

Golosov, V. N.: Erozionno-akkumulyativnye protsessi v rechnih basseinah osvoennih ravnin (Erosion and deposition processes in river basins of cultivated plains), Moscow: GEOS (Publ.), 296 pp., 2006 (in Russian).

Groisman, P. Y., Karl, T. R., Easterling, D. R., Knight, R. W., Jamason, P. F., Henessy, K. J., Suppiah, R., Page, C. M., Wibig, J., Fortuniak, K., Razuvaev, V. N., Douglas, A., Forland, E., and Zhai, P. M.: Changes in the probability of heavy precipitation: important indicators of climatic change, Climatic Change, 42, 243–285, doi:10.1023/A:1005432803188, 1999.

Larionov, G. A.: Vodnaya i vetrovaya eroziya: osnovnye zakonomernosti i kolichestvennaya otsenka (Water and wind erosion: main features and quantitative assessment), Moscow: MGU (Publ), 200 pp., 1993 (in Russian).

Lyuri, D. I., Goryachkin, S. V., Karavaeva, N. A., Denisenko, E. A., and Nefedova, T. G.: Dinamika selskohoziaystvennih zemel' v Rossii v XX veke i postagrogennoye vosstanovleniye rastitelnosti i pochv (Dynamics of agricultural lands in Russia in XX century and postagrogenic restoration of vegetation and soils), Moscow: GEOS (Publ.), 416 pp., 2010 (in Russian).

Madsen, H. and Thyregod, P.: Introduction to General and Generalized Linear Models. Chapman & Hall/CRC, 2011.

Nearing, M. A., Pruski, F. F., and O'Neal, M. R.: Expected climate change impacts on soil erosion rates: A review, J. Soil Water Conserv., 59, 43–50, 2004.

Park, H., Sherstiukov, A. B., Fedorov, A. N., Polyakov, I. V., and Walsh, J. E.: An observation-based assessment of the influences of air temperature and snow depth on soil temperature in Russia, Environ. Res. Lett., 9, 1–7, doi:10.1088/1748-9326/9/6/064026, 2014.

Petelko, A. I., Golosov, V. N., and Belyaev, V. R.: Experience of design of system of counter-erosion measures, Proceedings of the 10-th International Symposium on River Sedimentation, Moscow, 1, 311–316, 2007.

R Core Team: R: A language and environment for statistical computing, R Foundation for Statistical Computing, Vienna, Austria, available at: https://www.R-project.org, last access: February 2016.

Razuvayev, V. N., Apasova, E. G., Martuganov, R. A., Steurer, P., and Vose, R.: Daily Temperature and Precipitation Data for 223 U.S.S.R. Stations. ORNL/CDIAC, Numerical data package – 040, Oak Ridge National laboratory, Oak Ridge, Tennessee, USA, 1993.

Russian Research Institute of Hydrometeorological Information – World Data Center (RRIHI-WDS), Obninsk, Russia, available at: http://aisori.meteo.ru/ClimateR, last access: March 2016.

Rosstat – Russian Federal State statistics service (Rosstat-RFSSS), Moscow, Russia, available at: http://www.gks.ru, last access: March 2016.

Rysin, I. I., Grigoriev, I. I., Zaitseva, M. Yu., Golosov, V. N., and Sharifullin, A. G.: Long-term monitoring of gully erosion in Udmurt Republic, Proc. IAHS, in press, 2017.

Sidorchuk, A. Yu. and Golosov, V. N.: Erosion and sedimentation processes on the Russian plain, II: The history of erosion and sedimentation during the period of intensive agriculture, Hydrol. Process., 17, 3347–3358, doi:10.1002/hyp.1391, 2003.

Veselov, V. M.: PC archives of the State Data Holding and technology of their organization, Proceedings of RIHMI-WDC, 170, 16–30, 2002 (in Russian).

Zhao, G., Mu, X., Wen, Z., Wang, F., and Gao, P.: Soil erosion, conservation, and eco-environment changes in the Loess Plateau of China, Land Degrad. Dev., 24, 499–510, 2013.

Zolina, O.: Change in intense precipitation in Europe, in: Changes in Flood Risk in Europe, edited by: Kundzewicz, Z. W., IAHS Press, Wallingford, Oxfordshire, UK, Special issue no. 10, 97–120, 2012.

Yang, D. W., Kanae, S., Oki, T., Koike, T., and Musiake, K.: Global potential soil erosion with reference to land use and climate changes, Hydrol. Process., 17, 2913–2928, 2003.

Open Access
PIAHS

The potential for gamma-emitting radionuclides to contribute to an understanding of erosion processes in South Africa

Ian D. L. Foster[1,2], John Boardman[3,4], Adrian L. Collins[5], Ruth Copeland-Phillips[1], Nikolaus J. Kuhn[6], Tim M. Mighall[7], Simon Pulley[5], and Kate M. Rowntree[2]

[1]Department of Environmental and Geographical Sciences, University of Northampton, Northampton, NN2 6JD, UK
[2]Department of Geography, Rhodes University, Grahamstown 6140, Eastern Cape, South Africa
[3]Environmental Change Institute, Oxford Centre for the Environment, University of Oxford, Oxford OX1 3QY, UK
[4]Department of Environmental and Geographical Science, University of Cape Town, Rondebosch 7701, South Africa
[5]Sustainable Soils and Grassland Systems Department, Rothamsted Research, Okehampton, EX20 2SB, UK
[6]Department of Environmental Sciences, University of Basel, Klingelbergstr. 27, 4056 Basel, Switzerland
[7]Department of Geography & Environment, School of Geosciences, University of Aberdeen, Aberdeen, AB24 3UF, UK

Correspondence to: Ian D. L. Foster (ian.foster@northampton.ac.uk)

Published: 3 March 2017

Abstract. Several research projects undertaken by the authors and others over the last 14 years have used fallout and geogenic radionuclides for understanding erosion processes and sediment yield dynamics in South Africa over the last 100–200 years as European settlers colonised the interior plains and plateaux of the country and imported new livestock and farming techniques to the region. These projects have used two fallout radionuclides (^{210}Pb and ^{137}Cs) to date sediments accumulating in reservoirs, farm dams, wetlands, alluvial fans and floodouts and have used other fallout nuclides (^{7}Be) and long-lived geogenic radionuclides (e.g. ^{40}K, ^{235}U) as part of a composite fingerprint exploring contemporary sediment sources and changes to sources through time. While successful in many parts of the world, applying these techniques in Southern Africa has posed a number of challenges often not encountered elsewhere. Here we explore some of the benefits and challenges in using gamma-emitting radionuclides, especially ^{137}Cs, in these landscapes. Benefits include the potential for discriminating gully sidewall from topsoil sources, which has helped to identify contemporary gully systems as sediment conduits, rather than sources, and for providing a time-synchronous marker horizon in a range of sedimentary environments that has helped to develop robust chronologies. Challenges include the spatial variability in soil cover on steep rocky hillslopes, which is likely to challenge assumptions about the uniformity of initial fallout nuclide distribution, the paucity of stable (non-eroding) sites in order to estimate atmospheric fallout inventories, and the limited success of ^{210}Pb dating in some rapidly accumulating high altitude catchments where sediments often comprise significant amounts of sand and gravel. Despite these challenges we present evidence suggesting that the use of gamma-emitting radionuclides can make a significant contribution to our understanding of erosion processes and sediment yield dynamics. Future research highlighted in the conclusion will try to address current challenges and outline new projects established to address them more fully.

1 Introduction

A number of early Southern Hemisphere studies noted the potential for using fallout radionuclides, including ^{137}Cs, for dating sedimentary deposits and fingerprinting sediment sources e.g., Longmore et al. (1983) and Loughran et al. (1990) in Australia. Few studies were undertaken in southern Africa before the 21st Century. Exceptions include the work of Kulander and Strömquist (1989), Strömquist (1991) and Quine et al. (1999) in Lesotho and Owens and Walling (1996) in Zimbabwe. These studies often focused on the value of the technique for assessing erosion and soil redistribution rates.

South Africa, and especially the Eastern Cape, has witnessed high rates of erosion and extensive land degradation that appear in significant part to be associated with the arrival of European settlers from the early 19th Century onwards (Hoffmann and Ashwell, 2001; Boardman et al., 2015), yet few recent systematic studies have attempted to measure rates of erosion or fluvial sediment transport across the country. The most recent compilation of sediment transport data for Africa was published by Vanmaercke et al. (2014), including an accessible on-line database comprising sediment yield data from reservoir surveys and river monitoring. Figure 1 shows the frequency distribution of sediment yields for South Africa extracted from this database to which additional sources have been added (from Weaver, 1988 and Rowntree and Foster, 2012). The data confirm earlier published estimates that sediment yields generally range from <50 to $>900\,t\,km^{-2}\,yr^{-1}$, although the highest is a yield of over $11\,000\,t\,km^{-2}\,yr^{-1}$ not included in the Vanmaercke et al. (2014) dataset but measured over 3 years in the Roxeni Dam in the former Ciskei in the 1980s (Weaver, 1988). There are significantly higher yields in the rivers database (median $230\,t\,km^{-2}\,yr^{-1}$; $n = 38$) than in estimates from reservoir sedimentation data (median $84\,t\,km^{-2}\,yr^{-1}$; $n = 138$). This is surprising as the catchment areas are an order of magnitude greater for river than reservoir catchments (4350 and $477\,km^2$ respectively) and only three of the river estimates include bedload data. Current datasets are difficult to use as there are variable time periods in the reservoir and river datasets; very few yield estimates are available from former communal lands that are known to be much more heavily degraded (e.g. Hoffman and Ashwell, 2001) and no account is taken of changes in yield through time (most of the data are pre-21st Century). Our current knowledge of sediment yields is generally poor, yet we know that the lifetime of many reservoirs is under threat from extremely high sedimentation rates as demonstrated by Fig. 1. While models based on the Universal Soil Loss Equation (USLE) or the revised version (RUSLE) have been extensively used in South Africa (e.g. Le Roux et al., 2007), Rowntree (2012) criticised their use as scaling up to catchments introduces landscape storage elements that cannot be accounted for in these simple hillslope plot-based models.

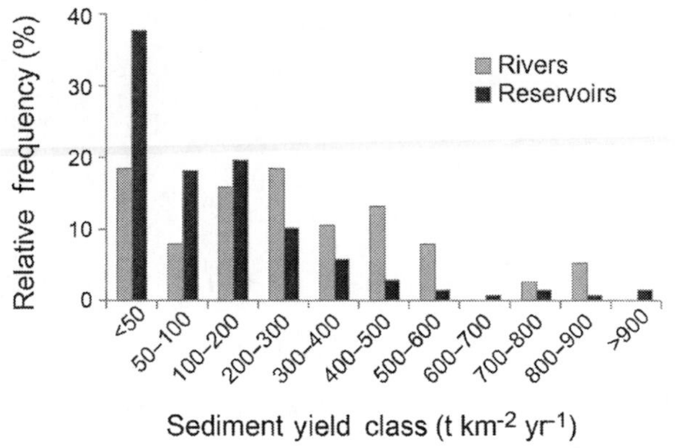

Figure 1. Frequency distribution of sediment yields in the South African Sediment Yield database with additional data (data from Vanmaercke et al., 2014; with additions).

Knowing rates of sediment transport or yield helps to understand the magnitude of the erosion problem but any effective solution to land degradation also requires that we know where the sediment has come from. We have used a range of techniques, including fallout radionuclides, in order to explore methods for reconstructing sediment yields over the last 1–2 centuries and for determining the provenance of that sediment. We have set this work within a conceptual and potential management framework that focuses on connectivity (e.g. Bracken et al., 2015) (see Foster et al., 2012 for detail).

2 Study area, materials and methods

Research has mostly focused on the Sneeuburg mountains and the plains of Camdeboo in the northern part of the Eastern Cape, South Africa. The highest mountain is the Kompasberg (2502 m a.s.l.). Geologically, the region is composed of mudstones and sandstones of the Karoo Supergroup (Permian to Triassic age) while dolerite dykes and sills (Jurassic age) dominate ridge tops and peaks. Soils on hillslopes are discontinuous but extensive valley fills, up to 5 m thick, exist in valley bottoms. The vegetation of the Sneeuberg is a mixture of grassland and shrub and is described in detail by Clark et al. (2009). Valley bottoms and valley side colluvium is often gullied to bedrock and, where badlands exist, A and often B horizons of soils have been stripped away. Annual rainfall ranges from 300 to 500 mm and is concentrated in the summer months from January through March (Foster et al., 2012).

Over the last 14 years we have quantified rates of sedimentation and/or reconstructed sediment yields in small farm dams using ^{210}Pb and ^{137}Cs to provide a chronology for each sediment stack. We have collected soil and sediment samples from contributing catchments and used a combination of particle size corrected geochemical, mineral magnetic, radionuclide and colour properties to determine the most likely or dominant contemporary sediment sources and to detect any changes in sediment sources through time (see Foster et al., 2012; Pulley and Rowntree, 2016). We have also measured erosion rates on badlands using long term erosion pin surveys (see Boardman et al., 2015).

The present paper focuses on the use of ^{137}Cs for both dating and fingerprinting sediment sources based on our experience of these catchments to date and attempts to explore some of the benefits and challenges involved. All radionuclide data reported here and elsewhere have been obtained using EG&G Ortec Hyper-Pure Ge detectors in a "well" configuration (40 mm deep; 11 mm diameter), located at Coventry and Northampton Universities and operated at liquid N temperatures. Samples were screened to $< 125\,\mu$m diameter (to match previous protocols; see Foster et al., 2007) by dry sieving prior to analysis and count times were typically in excess of 2 days in order to reduce counting errors to ca. 5 % or less. Spectra were processed using Ortec® Gammavision software and activities are expressed as mBq g^{-1}. Limits of detection are typically between 0.2 and 0.6 mBq g^{-1}.

3 Fallout and detection issues

Southern Hemisphere fallout was much lower than that of the Northern Hemisphere and peaked 1–2 years later than in the north, making detection of the first occurrence of ^{137}Cs in sedimentary deposits more challenging; Foster et al. (2012) now suggest that the first fallout in 1956 is probably undetectable using current detector geometry and count times and that the first occurrence in any sedimentary sequence is more likely to detect the cumulative fallout rise to between 1958 and 1960.

Measurement of potential sediment sources has revealed low (median ca. 2 mBq g^{-1}) and highly variable ($\pm$ ca. 2 mBq g^{-1} @ 5th and 95th percentile) ^{137}Cs activities in topsoils ($n > 150$). Many badland surfaces contain no detectable ^{137}Cs, although remnant undisturbed A horizons protected by individual Karoo shrubs within badlands often have detectable ^{137}Cs as well as excess ^{210}Pb and ^{7}Be (these appear to be in situ soils and not wind-derived). Gully sidewalls, excepting the upper 10–15 cm of the undisturbed profile, also contain no detectable fallout radionuclides. Cultivated soil profiles have lower activities and the ^{137}Cs extends to ca. 25 cm depth in the profile.

The high spatial variability in soil cover and depth on steep rocky hillslopes, coupled with the missing soil horizons in badlands, challenges any assumptions made about the likely uniformity of fallout nuclide distribution across these landscapes. Indeed, recent measurements of five 0–3 cm depth soil samples collected on a systematic grid from a small hillslope hollow on a sandstone lithology (ca. 0.5×0.5 m in area) revealed that ^{137}Cs could only be measured above limits of detection in 3 of the samples. Likewise, van der Waal (2015) detected ^{137}Cs in only 50 % of 28 surface soil samples collected from a small Drakensberg catchment. The combination of these factors suggest that the potential use of ^{137}Cs and other fallout radionuclides for estimating soil redistribution rates, using methods described by Walling and Foster (2016), are questionable at best.

An additional challenge in using fallout radionuclides for estimating soil redistribution rates is a paucity of stable (non-eroding) sites to obtain reference inventories. To date we have four inventories that are considered reliable; two cores in the Sneeuberg (averaging 41 ± 8 mBq cm^{-2}) and two cores on the plains of Camdeboo, receiving much lower rainfall (averaging 35 ± 6 mBq cm^{-2}) (Inventories are decay corrected to 2016).

At high altitude we have had limited success with ^{210}Pb dating in some rapidly accumulating farm dams where differences between supported and excess ^{210}Pb activities are within, or close to, the limits of analytical detection, invalidating ^{210}Pb dating models. ^{210}Pb dating in lower energy environments has, by contrast, been much more successful (see Foster et al., 2012).

4 ^{137}Cs in farm dam sediments

^{137}Cs inventories in seven farm dam sediment cores, all decay corrected to 2016, are variable and range from less than 190 to over 1300 mBq cm^{-2}, and are always significantly in excess of local reference inventories. Low inventories appear to be associated either with dams whose walls were breached, and remained unrepaired, or where sediment sources were dominated by badlands, where almost no ^{137}Cs is detectable in topsoil. Most non-breached dams have high post-1974 ^{137}Cs activities (when atmospheric fallout had almost ceased) suggesting that ^{137}Cs labelled topsoil continued to be delivered to all of the dams in significant quantities. Independent analysis using mineral magnetism, sediment colour and geochemistry (e.g. Foster et al., 2007) confirm this interpretation suggesting that gully sidewalls are not significant sediment sources but that the gullies themselves increase connectivity at the landscape scale by delivering sediment from more distal topsoil sources.

^{137}Cs profiles in the farm dams rarely mirror those of atmospheric fallout and have often proved equivocal in determining the peak in atmospheric fallout in 1965, unlike their Northern Hemisphere counterparts (see e.g. Walling and Foster, 2016). Lack of a consistent pattern appears to reflect variable delivery of excess ^{210}Pb to the dams from different contributing sources rather than variations in atmospheric fallout. Example profiles for two dams are given in Fig. 2 and show inconsistent patterns of activity down-core. What is notable, however, is that the activities in dam 37 (Fig. 2a) are much higher than in any topsoil or other potential sources yet measured in any of the catchments sampled (see Sect. 5 below for further discussion).

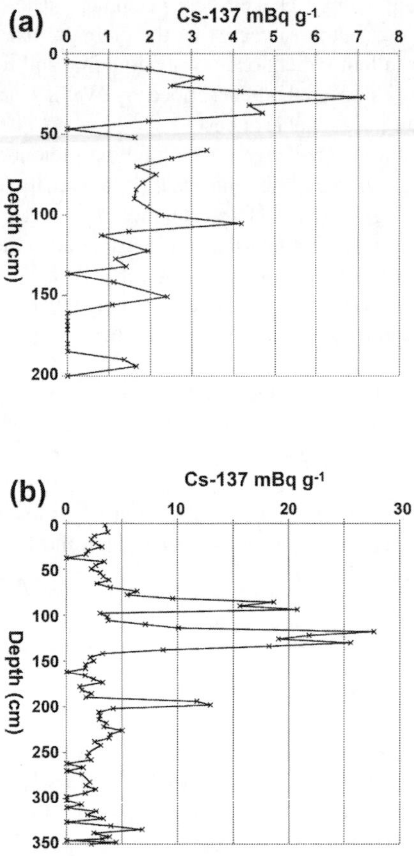

Figure 2. Downcore ^{137}Cs activities in 2 Farm Dams, Hartebeestfontein (Dam 94) and Good Hope (Dam 37) (The hiatus at ca. 55 cm depth in **(a)** indicates a period between a dam breach in 1974 and its subsequent repair in 1976).

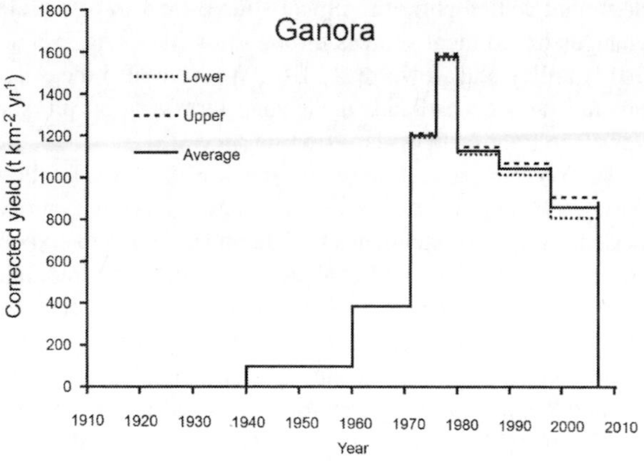

Figure 3. Long term sediment yields in the Ganora Catchment, Eastern Cape, reconstructed from dam sedimentation in Ganora Dam. Lower and upper values account for variable trap efficiencies (see Rowntree and Foster, 2012).

Similar radionuclide analyses, including the determination of excess ^{210}Pb profiles, have been undertaken on alluvial fan and floodout deposits (e.g. Foster et al., 2012) and although the 1965 fallout peak is again not clear, the first occurrence of ^{137}Cs across these landscapes in many depositional environments suggests that we can identify a common "time zero" against which rates of accumulation (and sediment yield in reservoirs) might be estimated over a known time period.

Detailed reconstructions of sediment yield have relied on ^{137}Cs and ^{210}Pb dating of single cores, supplemented by coring dates (or dates of breaching) and the known age of the dams, thereby bracketing the start and end of the depositional sequence. Additional chronological markers have been established from the presence of thin layers of fine gravels deposited in dams by high magnitude floods whose timeframe has been established from local rainfall records (e.g. Foster et al., 2007). Multiple sediment cores are required to quantify the spatial variability in deposition and estimate the volume of sediment deposited over different timescales. The chronology from the master core is then transferred to other cores using time-synchronous markers, such as magnetic susceptibility, which can be measured cheaply and rapidly, and has been shown to be a largely conservative tracer in dam sediments (Pulley et al., 2015). Detailed analysis of these sediment yield records is presented elsewhere (e.g. Foster et al., 2012) but clearly demonstrate that averaging erosion rates over long periods of time misses important information about the timing, magnitude and likely cause of any change in sediment yield (Fig. 3). This analysis suggests that the database shown in Fig. 1 is likely to be less robust as individual estimates of yield have been averaged over several decades in some cases and represent variable time periods and different durations, thereby limiting its value.

We recognise that there is a high cost in determining sediment yield histories using the reconstruction methodologies we have pioneered in South Africa. However, the fact that a common time line can be identified in sedimentary deposits provides an opportunity to obtain much more reliable information on spatial variations in sediment yield by targeting dams and reservoirs known to predate the mid-1950s and using ^{137}Cs to identify a common timeline in the deposited sediment. This means that a full ^{137}Cs profile is not needed once the maximum depth of ^{137}Cs activity has been determined but the approach would only provide a single yield estimate averaged over the last ca. 60 years. Multiple cores would still be required in order to undertake core correlation for the estimation of total sediment mass deposited since the late 1950s, but this could be based on cheaper analytical methods such as environmental magnetism (measurement facilities have been available in South Africa since 2009). Identifying 1958–1960 in lake and reservoir sediments using the fallout nuclide ^{137}Cs would allow a consistent estimate of sediment accumulation, and sediment yield, to be dated back to a common national baseline, making estimates between sites directly comparable. The advantage in this approach lies in significantly reducing the number of ^{137}Cs determinations likely to be needed at each site, making it possible to develop a large database at relatively low cost. Targeted full sediment yield reconstructions in carefully selected areas could take sediment yield estimates back to the early 20th century and would provide targets that could be used to demonstrate reductions needed to achieve a "baseline state" (see Foster et al., 2011).

5 The Good Hope Conundrum

It has already been noted that the downcore ^{137}Cs profile of Fig. 2b has some activities at least an order of magnitude higher than those measured in any potential source sample collected throughout the region. Even fractionation of five surface soil samples with high bulk activities to < 32 µm failed to produce ^{137}Cs activities significantly greater than ca. 12 mBq g^{-1}. The smallest 90th percentile of the particle size distribution of 90 samples analysed from the Good Hope dam sediment core was ca. 78 µm, with over 75 % of them > 150 µm. Furthermore, there was no statistically significant relationship between the down-core ^{137}Cs activity and particle size suggesting that this is not the major control on ^{137}Cs activities. Several questions therefore arise concerning the possible controls on these extremely high activities which are discussed briefly below.

In order to investigate other potential controls, such as the co-precipitation of ^{137}Cs with metal oxides or nutrients on particle surface coatings, concentrations of Fe, Mn and P were determined on microwave digested sediment samples using ICP-OES (Foster et al., 2007) but, again, no statistically significant correlations were found between metal or nutrient content and ^{137}Cs activity. Statistically significant correlations, with r^2 values generally above 90 %, were produced for several mineral magnetic susceptibility and remanence properties and, more surprisingly, with loss on ignition (LOI) (Fig. 4). In order to determine whether LOI was a true reflection of organic matter concentration in the sediment, and not C derived from bedrock or other sources, LOI results were compared with carbon analysis undertaken using a Carlo Erba C : N analyser at Rothamsted Research, Okehampton, Devon. Correlations between C and LOI were almost perfect (r^2 > 95 %), clearly demonstrating that LOI was a reliable measure of down-core trends in organic matter content in this core.

The above summary raises several questions about what is controlling the relationship between ^{137}Cs activity and mineral magnetic signatures and ^{137}Cs and organic matter content. These issues are currently under further investigation. The changes in mineral magnetic signatures are indicative of a change in sediment source in these catchments (see Foster et al., 2012; Pulley et al., 2015) and also suggestive of dolerite as being the potential source associated with the high magnetic signatures and ^{137}Cs activities. No known significant dolerite sources lie in close proximity to the Good Hope dam but further detailed reconnaissance surveys will be undertaken during future field visits in order to establish their presence. The second issue is more problematic as any of the myriad of methods that chemically remove organic matter from soil samples to establish if the ^{137}Cs is bound to it will likely attack the surface coatings of mineral grains where ^{137}Cs may also be bound (A. Horowitz, personal communication, 2016). In order to separate organic from mineral grains we are exploring a density-based separation technique using a high density liquid (Lithium Metatungstate [LMT] SG 2.95 mg L^{-1} cut with water to a SG of 1.3 mg L^{-1}) to physically separate the lighter organic matter from the heavier mineral soil using a non-aggressive liquid that is unlikely to chemically alter surface coatings (A. Horowitz,, personal communication, 2016). The detailed protocol is currently under development, using a preliminary version developed by Rothamsted Research, but after mixing ca 2 g freeze-dried sediment with ca. 30 mL of the 1.3 SG LMT, the suspension is centrifuged at 3000 rpm for 1 h and the LMT and organic matter poured off. Analysis will be undertaken on the residual organic and inorganic fractions separated.

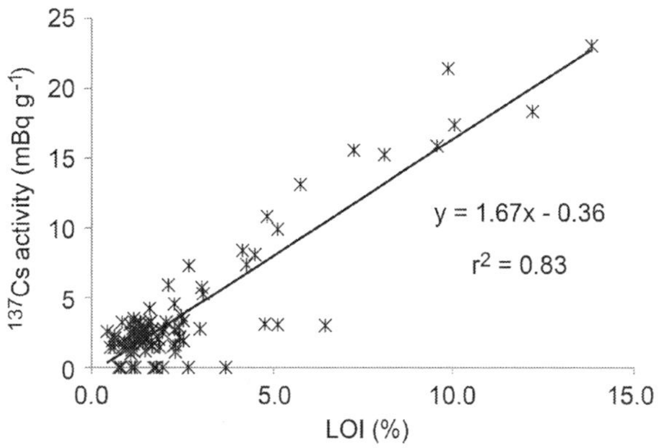

Figure 4. Relationship between LOI (%) and ^{137}Cs activity for the Dam 37 (Good Hope) sediment core.

6 Discussion and conclusion

The field data presented in this paper suggest that the measurement of ^{137}Cs and other fallout radionuclides offers significant potential for dating and fingerprinting sediment sources and for providing a possible baseline date in all depositional environments that effectively provides a time-synchronous marker horizon. Sediment accumulation rates and yields could be calculated from this baseline date in order to make comparisons at the local, regional and national scale using farm dams and reservoirs ranging in size from < 1 ha in area to large water supply reservoirs.

Interpreting the down-core changes in ^{137}Cs activities and inventories in most cores reflects changes in sediment source, changes in particle size and/or the relative dominance of ^{137}Cs labelled topsoil sources contributing sediment to the dams. The Good Hope dam ^{137}Cs profile remains problematic and further research is being undertaken to establish whether the high activities recorded in these sediments are associated with a change in sediment source, with the way in which ^{137}Cs is chemically bound to the deposited sediment, or with the way ^{137}Cs might be preferentially moved through the catchment with very fine sediment, thereby increasing measured activities.

7 Data availability

The majority of the data presented here are from an ongoing and, as yet, incomplete project. We will be making all data sets available on completion of the project and will publish the results and a link to the data sets which will probably be hosted by the University of Northampton. Please contact Ian D. L. Foster for further information.

Competing interests. The authors declare that they have no conflict of interest.

Acknowledgements. Many thanks to Paul Stroud (Northampton) for re-drawing the diagrams for publication. Research reported here has been supported by the National Research Foundation (NRF) of South Africa Grant No: 85358, and Rhodes University Council Research Grant 2015–2016.

References

Boardman, J., Favis-Mortlock, D. T., and Foster, I. D. L.: A 13-year record of erosion on badland sites in the Karoo, South Africa, Earth Surf. Proc. Land., 40, 1964–1981, 2015.

Bracken, L. J., Turnbull, L., Wainwright, J., and Bogaart, P.: Sediment connectivity: a framework for understanding sediment transfer at multiple scales, Earth Surf. Proc. Land., 40, 177–188, 2015.

Clark, V. R., Barker, N. P., and Mucina, L.: The Sneeuberg: a new centre of floristic endemism on the Great Escarpment, South Africa, S. Afr. J. Bot., 75, 196–238, 2009.

Foster, I. D. L., Boardman, J., and Keay-Bright, J.: Sediment tracing and environmental history for two small catchments, Karoo uplands, South Africa, Geomorphology, 90, 126–143, 2007.

Foster, I. D. L., Collins, A. L., Naden, P. S., Sear, D. A., Jones, J. I. and Zhang, Y.: The potential for paleolimnology to determine historic sediment delivery to rivers, J. Paleolimnol., 45, 287–306, 2011.

Foster, I. D. L., Rowntree, K., Boardman, J., and Mighall, T.: Changing sediment yield and sediment dynamics in the Karoo uplands, South Africa: post – European impacts, Land Degrad. Dev., 23, 508–522, 2012.

Hoffman, M. T. and Ashwell, A.: Nature Divided: Land Degradation in South Africa, UCT Press, Cape Town, 2001.

Kulander, L. and Strömquist, L.: Exploring the use of top-soil ^{137}Cs content as an indicator of sediment transfer rates in a small Lesotho catchment, Zeit. Geomorph., 33, 455–462, 1989.

Le Roux, J. J., Newby, T. S., and Sumner, P. D.: Monitoring soil erosion in South Africa at a regional scale: review and recommendations, S. Afr. J. Sci., 103, 329–335, 2007.

Longmore, M. E., O'Leary, B. M., Rose, C. W., and Chandica, A. L.: Mapping soil erosion and accumulation with the fallout isotope caesium-137, Aust. J. Soil Res., 21, 373–385, 1983.

Loughran, R. J., Campbell, B. L., and Elliott, G. L.: The calculation of net soil loss using Caesium-137, in: Soil Erosion on Agricultural Land, edited by: Boardman, J., Foster, I. D. L., and Dearing, J. A., Chichester, Wiley, 119–126, 1990.

Owens, P. N. and Walling, D. E.: Spatial variability of Caesium-137 Inventories at Reference Sites: an Example from Two Contrasting Sites in England and Zimbabwe, Appl. Radiat. Isotopes, 47, 699–707, 1996.

Pulley, S. and Rowntree, K. M.: The use of an ordinary colour scanner to fingerprint sediment sources in the South African Karoo, J. Environ. Manage., 165, 253–262, 2016.

Pulley, S., Rowntree, K. M., and Foster, I. D. L.: Conservation of Mineral Magnetic Signatures in farm dams sediments in the South African Karoo: the potential effects of particle size and post-depositional diagenesis, J. Soil. Sediment., 15, 2387–2397, 2015.

Quine, T. A., Walling, D. E., Chakela, Q. K., Mandringana, O. T., and Zhang, X.: rates and patterns of tillage and water erosion on terraces and contour strips: evidence from caesium-137 measurements, Catena, 36, 115–142, 1999.

Rowntree, K. M.: Fluvial Geomorphology, in: Southern African Geomorphology: Recent Trends and New Directions, edited by: Holmes, P. and Meadows, M. E., Bloemfontein, Sun Press, 85–140, 2012.

Rowntree, K. M. and Foster, I. D. L.: A reconstruction of historical changes in sediment sources, sediment transfer and sediment yield in a small, semi-arid Karoo catchment, South Africa. Zeit. Geomorph., 56, Suppl. 1, 87–100, 2012.

Strömquist, L. (Ed.): Monitoring soil loss at different observation levels; case studies of soil erosion in the Lesotho lowlands, Uppsala University, Department of Physical Geography UNGI Rapport Nr-74, 1991.

van der Waal, B. W.: Sediment connectivity in the upper Thina Catchment, Eastern Cape, South Africa, Unpublished PhD Thesis, Rhodes University, South Africa, 2015.

Vanmaercke, M., Poesen, J., Broeckx, J., and Nyssen, J.: Sediment yield in Africa, Earth Sci. Rev., 136, 350–368, 2014.

Walling, D. E. and Foster, I. D. L.: Using environmental radionuclides and sediment geochemistry for tracing and dating fine fluvial sediment, in: Tools in Fluvial Geomorphology 2nd edn., edited by: Kondolf, M. and Piegay, H., 183–209, Chichester, Wiley, 183–209, 2016.

Weaver, A.: Factors affecting the spatial variation in soil erosion in Ciskei: an initial assessment at the macroscale, in: Geomorphological Studies in Southern Africa, edited by: Dardis, F. G. and Moon, B. P., Rotterdam, A A Balkema, 215–228, 1988.

SMART – Sediment Mitigation Actions for the River Rother, UK

Jennine L. Evans[1], Ian Foster[1,2], John Boardman[3,4], and Naomi Holmes[5]

[1]Department of Environmental and Geographical Sciences, University of Northampton, Northampton, NN2 6JB, UK
[2]Geography Department, Rhodes University, Grahamstown 6140, Eastern Cape, South Africa
[3]Environmental Change Institute, University of Oxford, Oxford, OX1 3QY, UK
[4]Department of Environmental and Geographical Science, University of Cape Town, Rondebosch 7701, South Africa
[5]Department of the Natural and Built Environment, Sheffield Hallam University, Sheffield, S1 1WB, UK

Correspondence to: Jennine L. Evans (jennine.evans@northampton.ac.uk)

Published: 3 March 2017

Abstract. The River Rother, West Sussex, is suffering from excess sediment which is smothering the river bed gravels. This is thought to be exacerbating issues of pollution and degradation of ecosystems. This project aims to identify the severity, extent, possible causes and potential mitigation options available to reduce these pressures on the river. Data have been collected from ten sites to investigate the amount of sediment stored in the river bed gravels and cores obtained from four small reservoirs to establish rates of sedimentation and contribute to the construction of a temporal sediment budget over the last 50–100 years. Evidence suggests that tributary streams have more stored sediment per m^2 upstream of their confluence with the River Rother compared to the Rother itself. Reservoir core data indicate that sediment has accumulated more rapidly in the small reservoirs surrounded by mixed agricultural land compared to one surrounded by ancient woodland. These are preliminary results and work is continuing.

1 Introduction

In the UK, some rivers are failing to meet the criteria for achieving Good Ecological Status (GES) as set out in the European Union (EU) Water Framework Directive (WFD) (Defra, 2014; Official Journal of the European Communities, 2000). In many UK cases, including parts of the River Rother catchment, high sediment loads are the reason for failure to obtain GES (Environment Agency, 2016). Sear (1996) reported that the bed of the River Rother in West Sussex, UK had changed from one historically dominated by gravel, to one currently dominated by fine sand. While fine sand smothers the gravels there is an added issue of silt and clay transporting nutrients and pollutants into the river (Collins et al., 2009). These sediment pressures have been identified as key variables likely to prevent the attainment of GES in the Rother; the specific sources and causes of the problem have yet to be fully evaluated.

Excessive sediment in waterways causes negative direct impacts such as high turbidity and the siltation of river beds (Collins and Walling, 2007). High concentrations of sediment in the water column also have a number of indirect negative impacts on ecosystems, such as damaging fish gills and increasing eutrophication (Collins et al., 2010). Excessive sediment input is frequently caused by soil erosion on agricultural fields, although it has been suggested that channel banks can also be a major contributor (Pulley et al., 2015). The soil that erodes from fields can be delivered to waterways and roads where most negative impacts are felt (Boardman et al., 2003; Evans, 2010). This impacts society with financial costs to local authorities, the public and industries which depend on having a healthy river and good communication networks (Collins et al., 2010). Flood risk can be increased when soil erosion is increased. If soils are eroding into waterways it is usually due to an underlying issue with

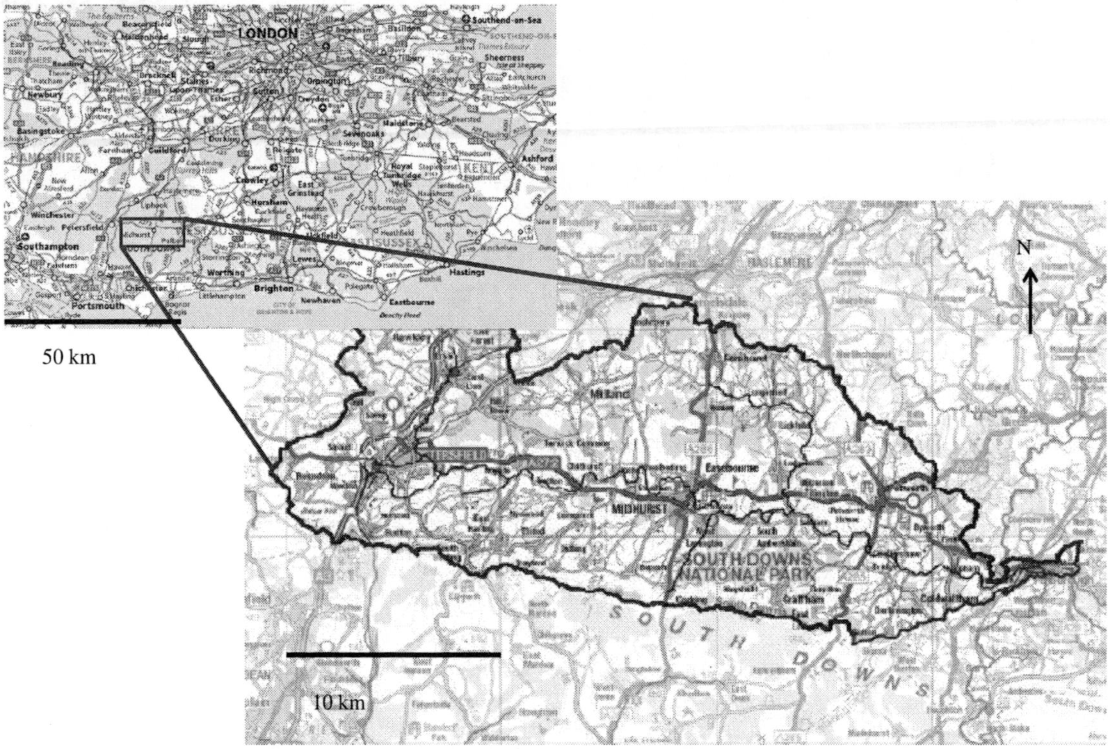

Figure 1. River Rother Catchment is located in the South Downs National Park with Midhurst at the centre (Edina Digimap, 2014; South Downs National Park Authority, 2015).

soil health. A reduced rate of water infiltration into the soil, or soil capping, increases the flashiness of surface runoff and increases sediment detachment and transport (Boardman and Favis-Mortlock, 2014). If the soil has a reduced infiltration capacity, excess water can create muddy flows that run across the land and into rivers, thereby increasing flood magnitude (Boardman and Favis-Mortlock, 2014).

1.1 Catchment description

The River Rother catchment is located in West Sussex/Hampshire in the South of England (Fig. 1). It drains 350 km^2 and the river itself is 52 km in length. The catchment marks the northern limit of the South Downs National Park. The two main tributaries of the Rother are the Lod Stream and the Hammer Stream, both northern (left bank) tributaries (Figs. 1 and 2).

Geologically, the area is underlain by Chalk and Lower Greensand of Cretaceous age and part of the Wealden group. The sandy soils developed on Cretaceous Greensand are naturally at high risk of erosion as they lack the cohesive bonding that silt and clay can provide (Evans, 1990). There are 21 soil associations within the Rother catchment (Evans, 1990). Seven of these soil associations are at "moderate" and "high" risk of soil erosion (Table 1); these cover ca. 64 % of the catchment. The topography is undulating with some locally steep areas and the altitude ranges from 0.4–250 m ODN. Average temperatures are between 6.2 and 15.1 °C and annual average rainfall is ~ 826 mm (Met Office, 2014).

Table 1. Soils at moderate (3) and high (4) risk of soil erosion within the Rother catchment (Evans, 1990).

Soil association No.	Soil association name	Erosion class
511g	COOMBE 2	3
554a	FRILFORD	4
571d	FYFIELD 1	4
571e	FYFIELD 2	4
571i	HARWELL	3
572k	BIGNOR	3
631d	SHIRRELL HEATH 2	3

Land use consists of arable land covering ca. 28 % of the catchment area with crops of winter wheat, maize, salad vegetables, potatoes and turnips which are all high erosion risk crops (Boardman and Favis-Mortlock, 2014; South Downs National Park Authority, 2015). Prior to the 1950s, the catchment was predominantly under permanent grazing. The steeper slopes further from the River Rother are mostly under grazing (ca. 36 %) and woodland (ca. 30 %). There are

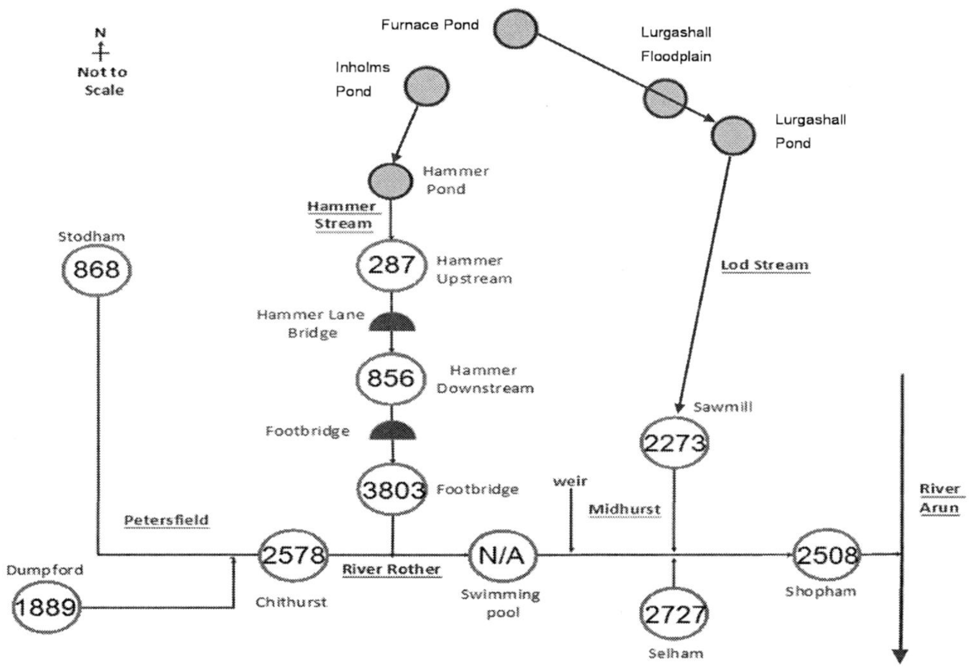

Figure 2. Locations of the monitoring sites in the Rother catchment (red circles) and coring sites (blue circles). Red circles display mean sediment storage at the monitoring sites ($g\,m^{-2}$).

a number of small villages and towns (ca. 4 %) but no major urban centres (South Downs National Park Authority, 2015).

1.2 Aims and objectives

This project aims to identify the severity, extent, possible causes and potential mitigation options available to reduce sediment pressures on the River Rother.

The objectives are;

1. To identify potential sediment sources and pathways for runoff and sediment transfer into the river system.

2. To reconstruct partial sediment budgets in the catchment over the last ca. 100 years.

3. To evaluate a range of sediment mitigation options that might be employed to control sediment delivery to rivers from sources and pathways identified in objectives 1 and 2.

2 Methods – river bed sediment sampling

In order to investigate the potential sources of sediment being carried by the River Rother and stored in, or smothering the substrate gravels, ten monitoring/sampling stations were selected. Stations include four along the main stem of the Rother, four on the Lod and Hammer streams, and two on small unnamed right bank tributaries, at Dumpford and Selham (Fig. 1).

At each site, a 10 L time integrated tube sampler was installed to collect suspended sediment at two month intervals (Phillips et al., 2000). At all sites except one, a river bed disturbance technique was used to take samples of the fine sediment deposited in the gravels (Collins and Walling, 2007). These samples will be compared with characteristics of potential sediment sources.

Coring

Sediment deposited in reservoirs (which are man-made, some for Medieval iron works) and floodplains can provide historical data as sediment transported by the rivers and streams settles at the bottom of receptors or on floodplains after a flood event (Pittam et al., 2008; Walling et al., 2008). Furnace Pond, Hammer Pond, Lurgashall Mill Pond and Inholms Pond (Fig. 1) were cored using either a Mackereth or a Russian Corer (Foster et al., 1998). Cores obtained reached depths of between 80 and 210 cm.

3 Preliminary results – river bed storage

Average bed sediment storage was estimated and results are shown in Fig. 2. In a study by Naden et al. (2016), the national range of stored sediment was found to lie between 6 and 4562 $g\,m^{-2}$ (median 181 $g\,m^{-2}$), with a similar range confirmed in an independent study in the UK by Collins and Walling (2007). It appears that all of the stored sediment estimates plotted in Fig. 2 are high compared with these studies, further suggesting that issues exist with stored sediment in this catchment irrespective of whether the sediment is also carrying any contaminants associated with it.

The site at Stodham has lower stored sediment compared to the rest of the monitoring sites indicating fewer sediment pressures in this headwater area, although sediment-associated pollution may possibly be a problem and is being investigated further. Most tributaries immediately upstream of their confluence with the Rother have high levels of stored sediment compared to Stodham and to the Hammer Stream immediately downstream of the Hammer Pond (Fig. 2). The river bed immediately downstream of the Hammer Pond had low fine sediment storage but the amount trapped rapidly increased immediately downstream of the Hammer Lane road bridge and exceeded average storage values for the main Rother channel. With the exception of the Hammer Stream, most tributaries appear to have similar amounts of sediment stored in channel beds.

3.1 Reservoir sediment coring

The Hammer Pond receives inflows from a mixed agricultural sub-catchment with arable, grazing and some woodland. Over 80 cm of sediment has accumulated in the Hammer Pond and Furnace Pond since 1954 (Fig. 3a, b) and over 150 cm in Lurgashall Mill Pond (Fig. 3c). These are some of the highest rates of sedimentation reported in the UK (Rose et al., 2010).

Inholms Pond is also in the Hammer Stream sub-catchment but is mostly forested with only a small area of permanent grazing land in the central area. Therefore, Inholms is being used as a reference catchment for comparison with other sites as it is relatively undisturbed. Sedimentation rates at Inholms Pond (Fig. 3d) are about half those of the Hammer and Furnace Ponds within the same time frame.

3.2 Conclusions and future work

This project is still in progress so conclusions are tentative. The river bed sediment storage data indicates that the Hammer Stream is accumulating more sediment per m^2 than the Rother itself. However identifying sources of the sediment in tributaries is key to understanding the sediment transport processes and accumulation in the Rother. Sediment source sampling is still in progress and will be analysed with a sediment unmixing model to determine provenance of the sediment in the waterways (Pulley et al., 2015). Sediment properties that will be used in this model include; particle size, mineral magnetism, radionuclides and sediment geochemistry.

The core data indicates that sediment accumulation is much higher in small reservoirs which receive sediment from mixed agricultural land as opposed to one surrounded by forest. The response of the floodplain upstream of Lurgashall Mill Pond is currently under investigation.

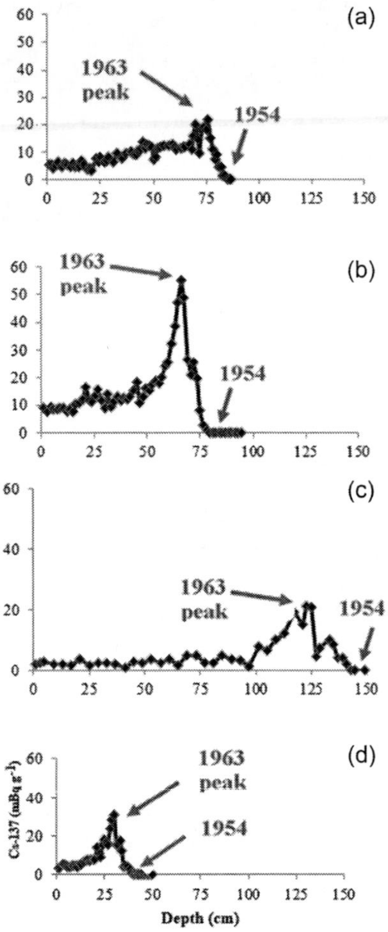

Figure 3. ^{137}Cs profiles; **(a)**, Furnace Pond; **(b)**, Hammer Pond; **(c)**, Lurgashall Mill Pond and **(d)**, Inholms Pond.

4 Data availability

The complete data set will be made public on the South Downs National Park (the funder) web site as an educational and research resource.

Competing interests. The authors declare that they have no conflict of interest.

Acknowledgements. We thank the South Downs National Park Authority (SDNPA) and the University of Northampton for funding the studentship associated with this project (JE) and the Arun and Rother Rivers Trust and Bruce Middleton (SDNPA) for helping secure site access and permissions to undertake this work.

References

Boardman, J. and Favis-Mortlock, D. T.: The significance of drilling date and crop cover with reference to soil erosion by water, with implications for mitigating erosion on agricultural land in South East England, Soil Use Manage., 30, 40–47, doi:10.1111/sum.12095, 2014.

Boardman, J., Evans, R., and Ford, J.: Muddy floods on the South Downs, southern England: problem and responses, Environ. Sci. Policy, 6, 69–83, doi:10.1016/S1462-9011(02)00125-9, 2003.

Collins, A. L. and Walling, D. E.: Fine-grained bed sediment storage within the main channel systems of the Frome and Piddle catchments, Dorset, UK, Hydrol. Process., 1459, 1448–1459, doi:10.1002/hyp, 2007.

Collins, A. L., McGonigle, D. F., Evans, R., Zhang, Y., Duethmann, D., and Gooday, R.: Emerging priorities in the management of diffuse pollution at catchment scale, Int. J. River Basin Manag., 7, 179–186, 2009.

Collins, A. L., Walling, D. E., McMellin, G. K., Zhang, Y., Gray, J., McGonigle, D., and Cherrington, R.: A preliminary investigation of the efficacy of riparian fencing schemes for reducing contributions from eroding channel banks to the siltation of salmonid spawning gravels across the south west UK, J. Environ. Manage., 91, 1341–1349, doi:10.1016/j.jenvman.2010.02.015, 2010.

Defra: Water Framework Directive implementation in England and Wales: new and updated standards to protect the water environment, May, 41 pp., available at: www.gov.uk/government/publications, 2014.

Edina Digimap: River Rother catchment: 50°59′8.31″ N, 0°44′21.82″ W, elevation 31M, Univ. Edinburgh, available at: http://digimap.edina.ac.uk/digimap/home, 2014.

Environment Agency: Catchment Data Explorer: Western Rother, available at: http://environment.data.gov.uk/catchment-planning/OperationalCatchment/3533, last access: 2 December 2016.

Evans, R.: Soils at risk of accelerated erosion in England and Wales, Soil Use Manage., 6, 125–131, doi:10.1111/j.1475-2743.1990.tb00821.x, 1990.

Evans, R.: Runoff and soil erosion in arable Britain: changes in perception and policy since 1945, Environ. Sci. Policy, 13, 141–149, doi:10.1016/j.envsci.2010.01.001, 2010.

Foster, I. D. L., Lees, J. A., Owens, P. N., and Walling, D. E.: Mineral magnetic characterization of sediment sources from an analysis of lake and floodplain sediments in the catchments of the Old Mill reservoir and Slapton Ley, South Devon, UK, Earth Surf. Proc. Land., 23, 685–703, doi:10.1002/(SICI)1096-9837(199808)23:8<685::AID-ESP873>3.0.CO;2-8, 1998.

Met Office: Midhurst Climate, available at: http://www.metoffice.gov.uk/public/weather/climate/gcp6p1zhh, 2014.

Naden, P. S., Murphy, J. F., Old, G. H., Newman, J., Scarlett, P., Harman, M., Duerdoth, C. P., Hawczak, A., Pretty, J. L., Arnold, A., Laizé, C., Hornby, D. D., Collins, A. L., Sear, D. A., and Jones, J. I.: Understanding the controls on deposited fine sediment in the streams of agricultural catchments, Sci. Total Environ., 547, 366–381, doi:10.1016/j.scitotenv.2015.12.079, 2016.

Official Journal of the European Communities: Directive 2000/60/EC of the European Parliament and of the Council of 23 October 2000; establishing a framework for Community action in the field of water policy, 2000.

Phillips, J. M., Russell, M. A., and Walling, D. E.: Time-integrated sampling of fluvial suspended sediment: A simple methodology for small catchments, Hydrol. Process., 14, 2589–2602, doi:10.1002/1099-1085(20001015)14:14<2589::AID-HYP94>3.0.CO;2-D, 2000.

Pittam, N. J., Foster, I. D. L., and Mighall, T. M.: An integrated lake-catchment approach for determining sediment source changes at Aqualate Mere, Central England, J. Paleolimnol., 42, 215–232, doi:10.1007/s10933-008-9272-9, 2008.

Pulley, S., Foster, I., and Antunes, P.: The uncertainties associated with sediment fingerprinting suspended and recently deposited fluvial sediment in the Nene river basin, Geomorphology, 228, 303–319, doi:10.1016/j.geomorph.2014.09.016, 2015.

Rose, N. L., Morley, D., Appleby, P. G., Battarbee, R. W., Alliksaar, T., Guilizzoni, P., Jeppesen, E., Korhola, A., and Punning, J. M.: Sediment accumulation rates in European lakes since AD 1850: Trends, reference conditions and exceedence, J. Paleolimnol., 45, 447–468, doi:10.1007/s10933-010-9424-6, 2010.

Sear, D. A.: Fine Sediment Accumulation in the River Rother, West Sussex, Environ. Agency, 1996.

South Downs National Park Authority: Land Use in the River Rother Catchment, Ord. Surv. 100050083, 2015.

Walling, D. E., Collins, A. L., and Stroud, R. W.: Tracing suspended sediment and particulate phosphorus sources in catchments, J. Hydrol., 350, 274–289, doi:10.1016/j.jhydrol.2007.10.047, 2008.

International Association of Hydrological Sciences
Centre for Ecology and Hydrology
Wallingford, Oxfordshire OX10 8BB
United Kingdom

ISBN 978-1-5108-4635-7
9 781510 846357

CURRAN ASSOCIATES INC.
proceedings
.com

26th Annual Saudi-Japan Symposium on Technology in Petroleum Refining and Petrochemicals 2016

Dhahran, Saudi Arabia
7 - 8 November 2016